Estimating Controls Systems for HVAC

Estimating Controls Systems for HVAC

A guide to pricing temperature controls systems for commercial buildings

Harry J. Edwards, Jr.
Controls Contractor, Richmond, Virginia

McGraw-Hill Book Company

New York St. Louis San Francisco Auckland
Bogotá Hamburg Johannesburg London
Madrid Mexico Montreal New Delhi
Panama Paris São Paulo Singapore
Sydney Tokyo Toronto

Library of Congress Cataloging-in-Publication Data

Edwards, Harry J.
Estimating controls systems for HVAC.

Includes index.
1. Commercial buildings--Heating and ventilation--Control--Estimates. 2. Commercial buildings--Air conditioning--Control--Estimates. I. Title.
Th7392.C65E38 1986 697 85-23124
ISBN 0-07-019094-1

1234567890 DOC/DOC 8932109876

ISBN 0-07-019094-1

The editors for this book were Joan Zseleczky and Susan B. West, the designer was Riverside Graphic Studio, and the production supervisor was Teresa F. Leaden. It was set in Baskerville by McGraw-Hill Computer Assisted Publishing.

Printed and bound by R. R. Donnelley & Sons Company.

For readers who are interested in copies of the forms appearing in Chapters 3, 9, 10, and 14, information regarding these forms can be obtained from the author at

1009 Baldwin Road
Richmond, VA 23229

DEDICATION

Estimating is a science and, as such, is subject to new discovery and improvement. From the takeoff for the estimate comes the framework within which the project will have to be installed. Thus, better methods of estimating actual contract costs are breakthroughs for contractors just as a discovery in the laboratory is a breakthrough for the scientist. This book on estimating control systems is dedicated to controls systems contractors, in the hope that they will profit from this work and will add their experience and personal discoveries to it, so that for them, controls systems estimating becomes a more exact science.

Contents

Preface

Temperature controls play a major part in maintaining comfortable conditions in a building. Their importance in a heating and air-conditioning installation is more apparent to the mechanical contractor than to the layperson, who may not appreciate the fact that automatic control systems are a major part of a mechanical contract. Frequently, the controls contract represents 10 percent of the cost of the mechanical work.

Bids for the control work might come from large national manufacturers or from franchised installers. National manufacturers maintain branch sales offices in cities throughout the country, they underwrite the operational costs of the offices, and they handle billing and collections. Franchised installers, on the other hand, are independent businesspeople who purchase from the large manufacturers and may even receive technical backup and supervision, but their offices are their own. Both national manufacturers and franchised installers specialize in the engineering and installation of automatic temperature control systems—pneumatic, electric, or a combination of both. Their bids relieve mechanical contractors (who are most often their customers) of the necessity of retaining personnel with the expertise to perform the control work. Instead, that portion of the work is subcontracted to temperature controls contractors, who handle the engineering, submittal, ordering, installation, checkout, and setup for a complete control system. This work is a specialty and requires a thorough knowledge of the fundamentals of temperature control and air-conditioning.

In recent years, major manufacturers have made their lines of controls available through wholesalers to the heating, ventilating, and air-conditioning trade. As a result many air-conditioning contractors have begun to examine the advantages of tackling temperature control installations themselves, and this has given rise to a generation of independent controls contractors who are competing in the open market for a share of the controls installation business. Many of these smaller contractors have good technical capabilities because of their experience with control systems in the field. This experience, coupled with whatever air-conditioning experience they may have, enables them to perform well in the areas of application; in addition, the literature that is shipped with the devices guides them in the proper methods of installation and calibration.

This broadening base of expertise has led to a need for good estimating methods that will produce accurate bids for temperature control work.

So much progress has been made in temperature control technology that courses dealing with automatic temperature controls are now routinely offered in many technical curriculums. From the days when a single thermostat was used to bring on heat, we have advanced to an era in which multiple thermostats within a single building are capable of delivering cooling or heating on demand, backed up by scores of relays, switches, and damper and valve actuators. Control sequences have become so intricate that cabinets of control devices are often necessary to effect day-night or summer-winter changeover or to present temperature and status displays as an aid to building management. Such an array of automatic equipment mandates the need to develop systematic and accurate estimating procedures that contractors can use to prepare a reliable estimate. Often contractors fail to recognize all the costs associated with control work because they use estimating methods that don't take all aspects of the installation into account.

Good estimating techniques are also necessary because manufacturers now make their control devices available for purchase over the counter. Without good estimating techniques, prices for control work could vary unreasonably on projects up for public bid. Therefore, a reliable estimating method should find wide acceptance in the field.

The control systems of tomorrow will be a hybrid of electronic and pneumatic systems. The use of electronic devices is steadily increasing because of the flexibility of electronic sensing, which allows paralleling of sensor input information into computer-based equipment for generating information for building management. These hybrid systems, are now becoming prevalent, and in many cases reduce the need for pneumatic work by requiring electronic instruments in place of pneumatic ones. The pneumatic portion of a control system will be its working end, where transducers within the control panels will deliver pneumatic signals to pneumatic control valves and to pneumatic damper actuators for modulating the flow of hot and chilled water and for modulating air flow within the ductwork. Programmable machine-operator interface modules will permit building owners and managers to do scheduling and to receive and display digital information about their systems.

The estimating method presented in this book is designed to permit electric devices, pneumatic devices, or a combination of both to be estimated using one set of estimating sheets. It permits you to take off and plan an entire project and arrive at a bid price, using in-house personnel for the installation if they have the necessary expertise and credentials. This method also allows you to approach either the pneumatic or the electric portion of a job separately and to subcontract one portion to others, if that is a better approach. An estimate produced according to this method is still valid if either the electrical or the pneumatic portion of the work is excluded. From labor-hour charts, you can select labor units for pneumatic devices or electric devices; this gives you latitude when making decisions about product mix based on your staff's capabilities.

As the emphasis shifts from one type of system to the other, such as from pneumatic to electronic, the estimating method presented in this book provides the flexibility needed to cross over from one craft's work to the other, all on the same estimate form.

Hand in hand with good estimating is good job management once the bid is won. Because each job won in competitive bidding is tight to begin with, you must carefully plan the expenditure of labor to avoid labor cost overruns, and therefore a labor planning sheet is presented for use in planning the installation of the job. Labor planning is part of the estimating process, and you are encouraged to look at your projected labor requirements. The labor planning sheet gives you an overview of labor-hours as you expect them to be needed on the job; then you can deploy your labor force as required. It's a tool you can't afford to be without.

In this book, I attempt to present a reliable method for estimating control work. It is not based on anyone else's method; rather, it is based on the need to recognize every segment of an automatic temperature control installation and to lead you to examine all phases of control work.

For over 20 years, I have worked in the temperature control field, and while I believe in the estimating principles presented in this book, I realize that no one method can meet everyone's needs. Therefore, you are encouraged to adapt and refine this method as you see fit, on the basis of your experience and your specific needs.

The need to manage energy efficiently has pointed up the need to modify controls in older systems in order to introduce system economies more in keeping with current-day design. Often the controls as installed are still quite serviceable, but certain retrofit work is necessary to modify the existing systems for greater operating efficiency. Consequently, you must be able to appraise an existing system and determine which additional pieces of control equipment are required to update it. The next-to-last chapter of this book (Chap. 19) focuses on what can be done to increase the effectiveness of older control systems and on how to estimate the retrofit work involved.

Although some engineering principles are presented in this book, the primary emphasis is on making the takeoff, preparing the estimate, and developing sufficient control diagrams that will ensure that the job is actually installed as you originally planned. The purpose of this book is to help you produce better estimates that will win jobs for you, while protecting your costs and enabling you to install high-quality control systems.

Where this book will fit the requirements of a course in the subject, I hope that it will prove beneficial in the class and later on the job.

I wish you all much success in the temperature control business. Good luck to those of you who will be managing your own businesses and taking the responsibility for the control work on projects that lie ahead. And thanks to the industry that has given me many years of happiness and excitement.

Harry J. Edwards, Jr.

Introduction

On first entering an equipment room, you might think that the work of the controls mechanic was very similar to that of an electrician if you had spotted the metal conduit which the controls mechanic had installed. However, if you looked more closely, you would see instrument-grade polyethylene tubing within this raceway, since metal conduit raceway is frequently used for housing polyethylene tubing for the compressed air when the tubing is exposed in a mechanical area such as an equipment room. If you looked in another part of the equipment room at the air handler, for example, you might think that a plumber had been at work, if you saw stretching across the air handler 1/4-in copper lines with soldered joints. Using such a run in copper can be a more economical alternative than extending the conduit system for only one or two tubes. In fact, pneumatic installation mechanics are more plumbers than electricians, since they're hired for their jobs out of the plumbers' and steamfitters' local and sometimes must install electrical metallic tubing when polyethylene tubing is to be installed within. But you still couldn't draw any conclusions, since the controls mechanic's expertise generally extends beyond the techniques of installation that you would see.

By training, controls mechanics are specialists. They recognize that installation is just the first phase of control work—that it constitutes only the rough-in of lines of communication for the controls. Once connected, each instrument reacts to changes in temperature, pressure, or relative humidity and initiates corrective action whenever there is a deviation from set point. Such reactions are precalculated, calibrated occurrences, determined by the commissioning of the device, which is done during the setup stage of control work. This is the work which sets the controls mechanic apart from an ordinary installer and commands special attention from the estimator, who must allocate time within the estimate for performing this work. Unlike an electric or plumbing fixture which is ready to operate once it has been installed, a controller must be assigned a set point and proportional band settings before it can perform as desired. Thus control system estimating embraces all the work to be done, including allotting the proper labor-hours for getting the job up and running.

Therefore, an acceptable method for estimating the cost of installing control systems must be able to take into account the various segments of labor to be found on any given job and must recognize that the intricacies of installation are time-consuming and have to be brought into the estimate as cost-accountable labor-hour requirements.

Changes in installation methods have been most dramatic for pneumatic systems. In the early part of this century, screwed pipe was frequently used for the air lines which served the devices. The devices themselves were considerably larger and heavier than those of today, and mounting took longer. When copper arrived on the scene as a viable alternative for the plumbing trades, it was selected as the more feasible form of piping for pneumatic systems as well. The devices featured 1/8-in female screwed ports for connection of the copper lines with compression-to-thread connectors. More recently, instrument-grade polyethylene tubing became the best choice for air lines since it is the least likely to leak. Because of the continuous manufacturing process that is used, long runs of resilient, airtight tubes are easily produced, and coupling points are minimized. The controls themselves are made with molded-in barbed connectors, which facilitate straight-on attachment to the tubing system. Estimating has evolved, too, into a more precise science, and the efficiencies of today's methods are reflected in pricing to the customer.

It is not unlikely then that this book will find its way into the reference libraries of many people in the heating, ventilating, and air-conditioning field. For as the cost of labor and of labor management increases, the trend toward supervised installations by major manufacturers will accelerate. These manufacturers will produce and distribute controls and furnish technical assistance; and in many cases, independent HVAC contractors will assume the responsibility for installation, along with smaller controls specialists' organizations. By then, almost all contractors in the HVAC field should be familiar with and using the estimating principles associated with temperature control work.

List of Abbreviations

A	auto
ac	alternating current
AFS	air flow switch
AHU	air-handling unit
AL	alarm
AM	auxiliary module
aux.	auxiliary
B	branch
bk.	break
C	common
C	controller
C1	cooling stage 1
C2	cooling stage 2
CC	cooling coil
cfm	cubic feet per minute
CK	check valve
CM	control module
CNW	condenser water
CNWP	condenser water pump
contr.	controller
CT	cooling tower
CWP	chilled water pump
CWR	chilled water return
CWS	chilled water supply
D	air damper
D	day
diff.	differential
dir. act.	direct-acting
DM	damper motor
ea.	each
EAD	exhaust air damper
EC	enthalpy controller
elect.	electric
EMT	electrical metallic tubing
EP	electro-pneumatic relay
equip.	equipment
ES	enthalpy sensor
ET	electric thermostat
F	fan
FMS	fan motor starter
FP/VAV	fan-powered variable air volume
FS	firestat
FZ	freezestat
H	hand
HL	high limit
HOA	hand-off-auto
HT	humidity transmitter
HVAC	heating, ventilating, and air-conditioning
HW	hot water
HWP	hot water pump
HWR	heating water return
HWS	heating water supply
HX	heat exchanger
ID	input and display module
ID	inside diameter
LL	low limit
loc.	located
M	main
M	damper actuator
mech.	mechanical
mk.	make
MP	minimum position switch
MS	motor starter
MSO	morning start-up optimizer

N	night
NA	not applicable
NC	normally closed
NCOA	normally closed to outside air
NO	normally open
NORA	normally open to return air
NT	night thermostat
O	off
OA	outside air
OAD	outside air damper
OAT	outside air thermostat
OD	outside diameter
OL	overload
OM	output module
OT	override timer
P	pump
Ⓟ	indicates a panel-mounted device
PE	pneumatic-electric relay
PHC	preheat coil
PM	power module
PMS	pump motor starter
pneu.	pneumatic
poly	polyethylene
PRV	pressure regulating valve
PRV	pressure relief valve
PS	pressure switch
qty.	quantity
R	relay
R	reset
RAD	return air damper
RAF	return air fan
RC	receiver controller
res.	restrictor
rev. act.	reverse-acting
RHC	reheat coil
rpm	revolutions per minute
S	signal
SAF	supply air fan
SP	static pressure sensor
SR	selector relay
SR	spring range
ST	space thermostat
stat	thermostat
T	thermostat
TC	temperature controller
TC	time clock
TG	temperature gauge
TH	temperature and humidity sensors
TIG	temperature-indicating gauge
TM	timer motor
TT	temperature transmitter
V	valve
VAV	variable air volume
vent.	ventilator
WFS	water flow switch
WU	warm-up
XF	transformer
XFMR	transformer

Estimating Controls Systems for HVAC

1

Control System Design and Specifications

In arriving at an understanding of the control systems marketing process, we would probably be well advised to examine the procedure generally employed in the construction of a commercial building. This procedure begins with the owner's retention of an architect who will accept the overall responsibility for bringing the structure into being. This responsibility includes preparing bidding documents, advertising for bids, awarding a contract to a general contractor, and supervising construction and completion of the project to the satisfaction of the owner. In preparing the bidding documents, that is, the plans and specifications, the architect uses professionals on staff as well as consulting engineers with special expertise in various areas of design. They include structural, electrical, and mechanical consultants, who handle plumbing and heating and air-conditioning. Because the consultants are specialists, they are able to provide greater proficiency in their areas of engineering application and design than could the architect alone. These consultants work daily in their particular fields and are in touch with the latest developments. The temperature control systems in a building are the responsibility of the mechanical engineers who design the heating and air-conditioning systems. Building the temperature control systems is specified as a part of the work of the air-conditioning contractor and is normally performed by a temperature controls subcontractor.

Often, consulting engineers look to their suppliers for the latest information about their products, and frequently will request the assistance of the suppliers in preparing specifications for work in their fields. And for years, as part of their sales strategy, major temperature control manufacturers have sought to furnish spec writing assistance to consultants. Consulting engineers often gladly involve controls sales representatives in preparing specifications when the consultants are assured that the controls rep (1) is genuinely interested in being of help to the engineer and (2) can provide the best controls for the job at a fair price using system design that is practical and efficient. A controls rep who has earned the respect of a consultant should be careful not to lose it through poor design or overpricing. The controls field is quite competitive, and contractors recognize the value of enjoying the approval of the mechanical design consultant. All those who can be of real assistance will surely be invited to participate in spec writing.

THE TEMPERATURE CONTROLS SPECIFICATION

The temperature controls specification usually begins with an opening paragraph that describes the scope of the temperature control work and names acceptable manufacturers. Subsequent paragraphs allocate responsibility for electric wiring, whether the control system is pneumatic or electric, or a combination of both, since pneumatic controls also require electric wiring where they interface with motor starters, pilot lights, or other electric apparatus. Additionally, the air compressor for the pneumatic system must be wired.

A further paragraph details the type of tubing to be used on the project. This paragraph is followed, perhaps, by one on automatic control dampers and then one on automatic control valves. The installation of dampers and valves is part of the work of the mechanical contractor. Automatic valves and dampers are furnished by the controls subcontractor and are installed by the mechanical contractor. Dampers are mounted in the ductwork or in air handlers, and this craft work is handled by either the mechanical contractor or the sheet metal contractor (a subcontractor to the mechanical contractor). Valves are piped in by the pipefitters working for the mechanical contractor. These installations are made under the supervision of the controls contractor.

Next, the specification discusses the control devices which will be used on the job, listing such things as temperature ranges, capillary lengths, switching points, electrical characteristics for interface devices, enclosures, mounting, and mode of control. A paragraph may be included which speaks to checkout, system start-up and commissioning, owner orientation, and installation warranty. It is customary to specify a 1-year warranty period during which malfunctions are to be corrected by the controls contractor at no cost to the owner; routine maintenance is not covered by the warranty. Routine maintenance is reviewed at the orientation session and is to be provided by the owner in accordance with maintenance and repair sheets.

Sequence of Operation

At this point in the controls specification, the sequence of operation for the various pieces of mechanical equipment under control is given. Included in this section are details of how the heating, ventilating, and air-conditioning (HVAC) equipment reacts to temperature changes both within the unit and out in space. In space, the room thermostat may dictate the delivery of hot or cool air from the unit, or remote terminal units might select cool air delivered at a rapid rate for cooling or reheated air delivered at a slower rate for heating. Unit heaters, fan coil units, converters, baseboard finned-tube radiation, condenser water temperature, variable-volume units—all these and more—are described in terms of their response to thermal demands within the space being served subject to the direction of the temperature controls. It is from the sequence of operation that the controls contractor will later draw control diagrams to aid in the takeoff.

Thus the temperature controls specification is developed by the consulting engineer with whatever assistance suppliers can provide to delineate the reaction which is to be expected in response to changing temperatures. The specification is the road map for the preparation of bids by controls subcontractors for the controls work. It is the basis for the in-depth discussion of controls estimating in this book. By definition, the controls sequence of operation represents the benchmarks to be observed when the controls encounter given conditions of temperature change.

Placement of Controls and Limits

Good controls designers can apply their experience in the field to the proper placement of controls within the air handler and to the application of low and high limits. Controlling the discharge temperature of a direct expansion cooling coil is impractical, as the sudden change in the temperature of the air leaving the cooling coil produces short cycling of the air-cooled condensing unit serving the coil. A far more practical approach that is generally approved by air-conditioning equipment manufacturers is to use a return air thermostat to activate the mechanical cooling equipment and room thermostats to add reheat if subcooling occurs, or to accept the signal of the warmest zone thermostat to energize the condensing unit, allowing the compressor and its capacity controls to regulate condensing unit output to match coil load.

Unit Shutdown

Annunciation of unit shutdown due to actuation of a safety control is an appreciated design feature from the owner's standpoint. The tripping of a firestat or freezestat without alerting building personnel could result in unnecessary temperature deviations in the area served. And misapplication of freezestats produces significant problems for building managers during the cold winter months as units are continually shut down from lockout when excessively cold temperatures within an air handler strike the freezestat capillaries without the benefit of mixed air low limits or preheat coils.

RESPONSIBILITIES OF THE CONTROLS CONTRACTOR

Experienced controls installers are mindful of an extensive list of areas to be approached with caution, which makes them valuable consultants to the mechanical designer. It is with good reason, then, that people in the HVAC design field call upon the controls contractors for assistance in writing specifications. And those contractors who accept this challenge are presented with the responsibility of applying controls effectively and economically, with an eye toward avoiding nuisance situations that can develop through shortsighted selection and application of devices. This responsibility is part of the effective marketing of controls. As controls systems representatives, controls contractors can play an important part in the design process and the better the job done by the designer before the bid, the better the job the controls contractor can do in pricing and quoting.

I feel that the controls contractor has an obligation to point out to the mechanical designer any area in an application the designer is planning which the contractor knows through experience may present a problem. It is as a team that the controls contractor and the mechanical designer can develop the most trouble-free system for the owner. Working together in this way is sure to build goodwill for both the controls contractor and the designer and to strengthen future ties. It is by acting in the best interests of all concerned that the success of the job is made more certain, and it is likely that, upon completion of the design, a good system will be sold quickly and accepted by a satisfied owner.

I'm sure you can think of any number of areas in years past where experience has taught you that doing something another way would have been better. Such is the nature of hindsight. But hindsight is of no benefit if you cannot profit from it in future situations. Thinking back over jobs you have completed will bring to mind certain problems that arose during the progress of the work, and how you solved them. You can use this experience to help in weeding out weaknesses in projects in the design stage.

VAV SYSTEMS

One area that can be a problem concerns variable air volume (VAV) systems, which are subject to nighttime shutdown, particularly where low limits are maintained by perimeter heat. As estimators, controls contractors need to appreciate the difficulties inherent in VAV systems which can cause occupant discomfort and owner confusion if room temperatures are inconsistent. Invariably, interior spaces are served with VAV boxes which do not use reheat coils. Without reheat coils, the only options in core offices are cooling and no cooling, because a fall in temperature causes the controlling thermostat to drive the VAV box output to minimum. It is at this minimum point that control ceases should the room temperature continue to fall. If the occupants are out of the room and the lights are off, and the room temperature is low enough so that the box is positioned for minimum output, a minimum amount of 55°F air is constantly being introduced into this office, and subcooling may occur in the absence of a heat load generated by people and lights. Consequently, when the occupants reenter the office, they will find it too cool, and perhaps will turn up the thermostat—but to no avail. The box was already set for its lowest air delivery as a result of a low room temperature. Warm-up of the room depends upon lights and people creating a load through the release of Btu's. So the occupants must settle for a delayed rise in temperature until their own body temperature, physical activity, and lights create a cooling load which will raise the room temperature to acceptable levels. It is, however, sometimes difficult for people to be content with anything less than an immediate response when they reset their thermostat, but in a typical interior space served with a simple VAV box without a reheat coil, there is little alternative, short of system modification, to relying upon internal loads to create a rise in room temperatures. Conditions are better, usually, if 100 percent shutoff is obtainable within the VAV box.

In a similar situation, VAV systems in which air handlers are shut down overnight frequently rely upon perimeter radiation for maintaining reduced nighttime set points. As interior temperatures fall to 55 or 50°F, night setback thermostats in perimeter offices control radiation to maintain perimeter temperatures. These same day-night thermostats are remotely indexed back to higher daytime temperatures when normal occupancy time approaches. During the day, the thermostats sequence the downward modulation of perimeter radiation as the output of the room VAV box increases with a rise in temperature. Thus, in the morning, at the time of indexing back to daytime conditions, the perimeter controls are able to add heat via radiation while VAV boxes serving these areas are held at minimum output. This is not so, however, with the interior or core offices. Usually, these offices are controlled by single-temperature thermostats, which have no ability to add heat—setting these thermostats back at night would be useless with the fan off and no radiation available. Upon a return to daytime conditions, the chilly core offices remain so, as heat pickup must migrate toward the core from the perimeter areas, or be obtained from the normal output of body heat from personnel and lighting. Thus, core offices sometimes remain uncomfortably cool until late morning or noon when normal occupancy has had a chance to raise the temperature. The problem of interior office spaces that are too cool in the morning can frequently be attributed to a design which called for cooling-only VAV units to serve them.

A controls contractor may want to suggest that the consultant utilize the air handler to achieve morning warm-up. In a classic warm-up application, a return air thermostat set at perhaps 65°F assumes control of a heating coil within the air handler *during warm-up only*. During this warm-up period, control is temporarily removed from the thermostats, and all VAV boxes are automatically overcalled and positioned for full air flow. Perimeter radiation, however, remains under the control of the thermostats in the perimeter offices, and the radiation output diminishes as the space temperature rises as a result of the action of the return air warm-up thermostat. Once the return air reaches 65°F, all thermostats are again placed in control, and perimeter as well as interior spaces are ready for the arrival of personnel. The warm-up period is often a function of an optimizing type of time clock which lengthens the period of warm-up as dictated by inside and outside temperature differentials. Once the unit comes out of warm-up, and normal discharge temperatures of 55 to 60°F are reestablished, the heating coil is shut down, thus allowing discharge temperature to resume control of the system. It can be concluded then that a warm-up cycle in conjunction with a VAV system is frequently an advisable combination.

VAV systems are energy savers. As heat-satisfied spaces cause their VAV terminal units to throttle back on air flow, static sensors reposition the capacity control dampers in the air-handling unit's fan to take in less air, thereby easing the load on the fan and its electric motor, resulting in a savings of electric energy. Reduced air flow over the air-handling unit's chilled water coil causes a fall in the chilled water return temperature, and the water chiller begins to unload as well. Selection of VAV systems for commercial space conditioning is founded on sound energy conservation principles, but application of controls and strategic placement of heating coils should also warrant some consideration. The institution of a morning warm-up period in cold weather can play an important part in assuring comfortable core area offices for arriving morning workers.

It is easy to see that control system design and specifications can often be as much the responsibility of the controls contractors as that of the consulting engineer, for controls contractors have an obligation to relate their experiences in the field so that the consulting engineers can benefit from those experiences. There is no substitute for people working together when it comes to attaining the highest quality with economy.

2

Developing the Temperature Controls Takeoff

The first step in the submission of a proposal is the preparation of an estimate. Formulating an idea about project cost is necessary whether the proposal takes on the formality of a written presentation or merely consists of a ballpark figure for budget purposes. In each case, a review of material and labor is conducted, with an eye toward establishing a cost basis from which future purchase orders and work-force plans will evolve. Even the simplest of jobs requires coming up with an idea about the number of devices and labor-hours necessary for producing a working control system that will benefit the owner.

PERFORMANCE SPECIFICATIONS

A basic difference between different approaches to controls system estimating is how the hardware that will carry out the sequence of operation for achieving space temperature control is selected. One approach uses a performance specification, which merely states the sequence of operation and leaves it to the controls specialist to draw up a control diagram and choose hardware that will carry out these performance requirements. Another approach is to use a technique specification, which describes the hardware and frequently shows it in schematic form. The advantages of using a performance specification rather than a technique specification are that a performance specification allows for greater flexibility in the selection of devices and may produce more competitive bids, since individual controls specialists may use different approaches and equipment to construct a custom system to fit the purchaser's needs. Preferential treatment of any one controls supplier is avoided, since equal latitude in device selection is afforded to all bidders. By not specifying technique or making device selection in the bid documents, the product mix will not be slanted in favor of any one supplier.

PERFORMANCE SPECIFICATION FOR AN AIR HANDLER

In the sample paragraph below, from a mock performance specification, you can see that limitation to specific hardware is not the case. Rather, bidders are free to apply their controls as they see best in producing a control sequence to satisfy what is specified. Consider the following operational description as might be written for an air-handling unit that uses fresh air and chilled water for cooling and hot water for heating.

> The unit shall be automatically started at a predetermined time each morning. Upon start-up, the fresh air damper shall assume a minimum position to ensure the introduction of a minimum amount of fresh air necessary to satisfy area requirements. When the space temperature is below the 55°F set point of the dual heating-cooling room thermostat, the heating coil valve shall be positioned for full coil flow, and the unit shall be on full heat. As the space temperature approaches the room thermostat's heating set point, the heating coil valve shall be modulated to diminish hot water flow to the coil. On a continued rise in temperature, and after hot water flow to the heating coil is stopped, no further reaction shall occur at the air handler until and unless the space temperature rises to within the throttling range of the cooling thermostat's set point. When the space temperature approaches the cooling thermostat's set point of 78°F, the fresh air damper shall be modulated open for free cooling as the return air damper is modulated closed a proportionate amount, allowing the introduction of greater quantities of fresh air for natural cooling. Should the space temperature continue to rise after 100 percent fresh air has been introduced, the cooling coil valve shall be positioned for as much flow of chilled water through the cooling coil as is required to maintain the space temperature. On a fall in the space temperature, the reverse of the above sequence shall occur.
>
> In the event that the mixed air temperature entering the unit falls below 55°F, the room thermostat shall be overcalled, and the fresh air damper modulated back enough to limit the mixed air temperature to a low of 55°F. Should the fresh air temperature rise above 60°F, where it is no longer suitable for natural cooling, the room thermostat shall be overcalled, and the dampers returned to their preset minimum position during operation of the air-handling unit.
>
> If for any reason the mixed air temperature falls to 35°F, the return air temperature rises to 125°F, or the discharge air temperature rises to 200°F, the fan in the unit shall be shut down and an audible alarm sounded in the building engineer's office. Whenever the fan is stopped for any reason, the fresh air damper shall fully close while the return air damper shall fully open.
>
> At night and on weekends, the air handler shall operate as a unit heater, with the fan being cycled on when the space temperature falls to an unoccupied setting. No fresh air shall be introduced during unoccupied periods, and the heating coil control valve shall be positioned for full coil flow.

The solution to this operational description can be approached pneumatically or electrically.

Pneumatic Solution

The first takeoff discussed will be pneumatic. Figure 2-1 is a pneumatic controls takeoff sketch showing the necessary devices in their sequential arrangement. It will be helpful to refer to Fig. 2-1 during the discussion about equipment selection.

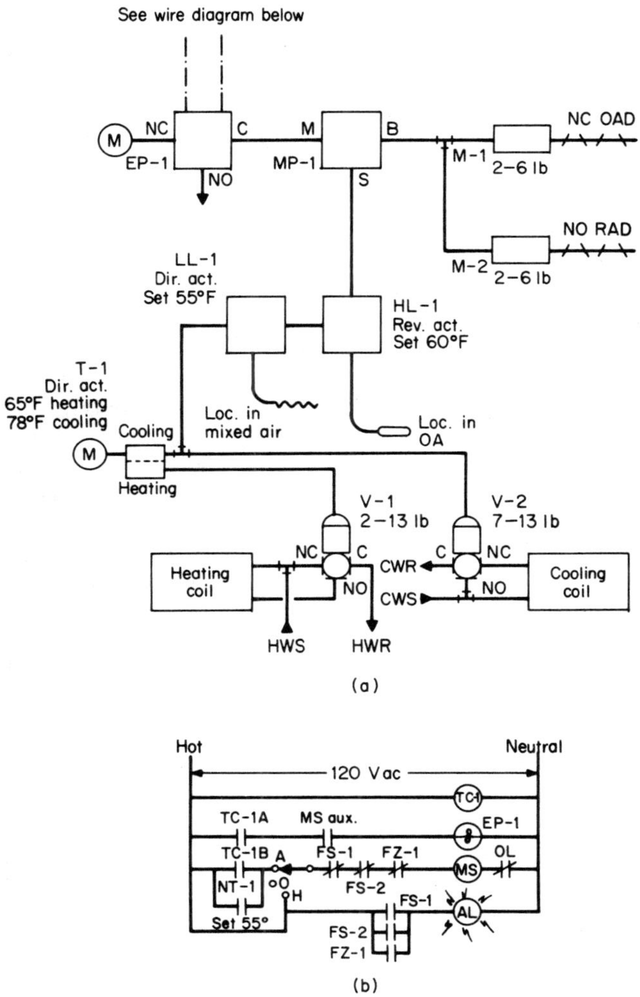

Figure 2-1 Heating-cooling air-handling-unit pneumatic controls takeoff sketch. (*a*) Pneumatic diagram; (*b*) wire diagram.

Automatic start-up is handled using time clock TC-1. The time clock is wired into the fan motor starter holding coil through the high-limit and low-limit safeties, firestats FS-1 and FS-2 and freezestat FZ-1, to initiate start-up. Notice that fan control contact TC-1B can be overridden by a separate night thermostat set at 55°F but ventilation control contact TC-1A cannot be. Lockout of fresh air must be ensured when the unit fan is cycled on for maintenance of reduced nighttime temperatures. Next a fixed minimum amount of fresh air is provided through the use of minimum position switch MP-1. When air from the electro-pneumatic relay is released by EP-1, MP-1 passes just enough control air to cause a minimum amount of damper motor movement, which allows introduction of a little fresh air whenever the unit is in operation during occupied periods.

Next, the sample specification covers space temperature, and so the need for a room thermostat is apparent. Additionally, since set points of 65°F for heating and 78°F for cooling are cited, the room thermostat must have a deadband to facilitate an interim period between set points when no heating is required and no cooling is taking place. A good choice is a space thermostat with separate branch lines that features two independent heating-cooling thermostats under a common cover. Such thermostats, known by different names, allow the space temperature to float between its respective heating-cooling set points without using any energy for heating or cooling, thereby effecting system energy savings. The branch line from the heating portion of the thermostat goes only to the heating coil valve V-1, while the branch line from the cooling portion of the thermostat goes only to the dampers and cooling coil chilled water valve V-2. As the set point of the heating thermostat is exceeded, the heating control valve goes into the coil bypass position. Once the valve is in this position, the heating portion of the thermostat can do nothing more, even if the temperature in the room continues to rise. Temperature control beyond this point must come from the cooling portion of the room thermostat. It modulates the intake dampers M-1 and M-2 for up to 100 percent fresh air, and if the space temperature still rises, positions the chilled water control valve for flow through the chilled water coil.

Next, the sample specification covers limits, and so provision must be made to prevent the air handler from taking in air which is too cold or too hot. To prevent the intake of air that is too cold, mixed air low-limit controller LL-1 is added to retard the pneumatic signal to the damper actuators as much as may be required to maintain the mixed air entering the unit at a minimum of 55°F. On the other hand, outside air temperature which is too high no longer helps to cool the spaces served, but rather begins to add to the cooling load. To prevent the intake of air that is too hot, fresh air high-limit controller HL-1 is interposed into the pneumatic signal to damper actuators M-1 and M-2 in series with LL-1. Now HL-1 too can override room thermostat T-1 when necessary to prevent the introduction of excessive amounts of hot, humid air in the summertime.

To fulfill the requirements for fan shutdown when air temperature extremes are encountered in the return air, mixed air, or discharge air, two double-pole firestats and a double-pole freezestat are chosen. All are wired with their normally closed contacts in series to deenergize the motor starter of the air handler should their temperature set points be reached. Note that double-pole safeties are selected, each with a normally closed and a normally open pole. The normally open contacts of the firestats and freezestat are wired in parallel to an audible alarm in the building engineer's office so that any of the three safeties can trigger the alarm independently of the others.

Nighttime temperature control in the winter is transferred to the night thermostat when fan control contact TC-1B opens. NT-1, which is set at 55°F, may restart the air-handling unit fan motor when the space temperature falls to its set point, and all fresh air will be held out, since EP-1 does not energize during unoccupied periods because ventilation control contact TC-1A remains open.

The takeoff sketch for the pneumatic system, shown in Fig. 2-1, has evolved from interpretation of the HVAC designer's written description of system operation. Component selection is left up to the individual bidders, whose job is to estimate the controls necessary to meet the performance called for in the operational sequence. And by drawing out the system, complete with piping, bidders can ensure the workability of the devices they have chosen, since control components are pictured in their working relationship with one another.

Electrical Solution

This same control system for the air-handling unit could have been approached electrically. A look at an electrical control schematic, where the system is developed with purely electric devices, permits comparison of the two approaches. When significant numbers of controls are involved, pneumatic controls usually hold the edge over electric controls in terms of cost savings, because once the initial cost of the air compressing outfit is offset, the economies of installing single plastic tubes over metal raceways with two wires become obvious. Figure 2-2 is an electric controls takeoff sketch of comparable electric controls that will give the same mode of control as the pneumatic system just discussed. The guide is the same performance specification from which the pneumatic system was developed, and care has been taken to serve each segment of the sequence of operation as thoroughly as before. Whether pneumatic or electric, each control will ultimately be costed out on an estimate sheet, including labor-hours and tubing or wire. Whereas the pneumatic takeoff sketch (Fig. 2-1) showed tubes in bundles, single-line tubing, or tubing in conduit, the electrical takeoff sketch (Fig. 2-2) shows electrical metallic tubing (EMT) and wire exclusively. Rigid galvanized conduit could have been selected, but its use in ordinary commercial temperature control work is rare, so EMT is the standard raceway here. Of course, electricians do the installation and set and connect the control devices. In this case, electricians perform the complete installation, including checkout and calibration and setup. Independent controls contractors probably would have mechanics who are as comfortable with wire as they are with tubes and have the necessary credentials to perform the electrical work as well.

The similarities between the two systems in Figs. 2-1 and 2-2 are obvious. Both figures show damper actuators M-1 and M-2, except in Fig. 2-2 they are electric actuators. Minimum position switch MP-1 is electric also, interposed into the modulating control circuit to introduce minimum actuator movement in the absence of an overriding signal from the C1 portion of cooling thermostat T-1 through low-limit controller LL-1 and high-limit controller HL-1. A

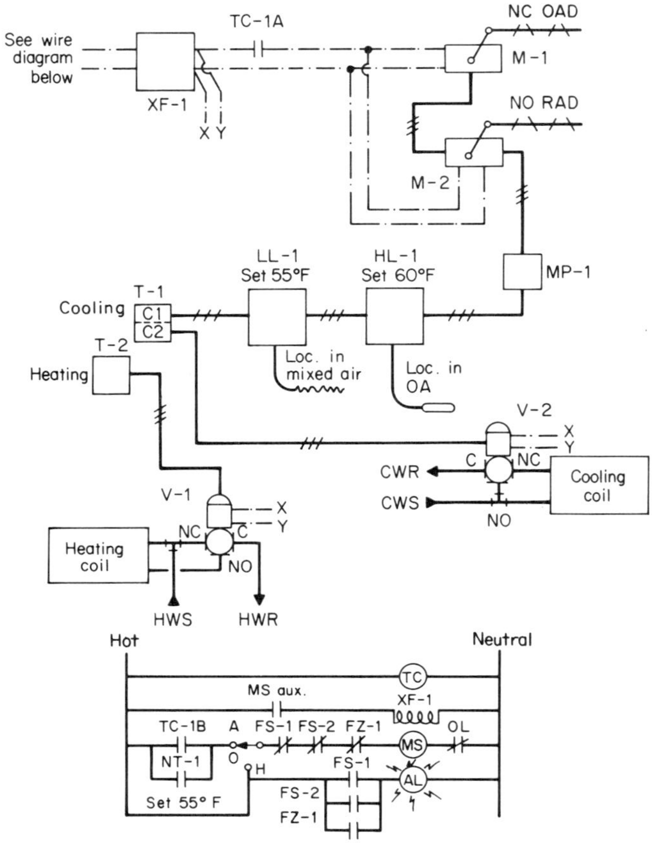

Figure 2-2 Heating-cooling air-handling-unit electric controls takeoff sketch.

separate heating thermostat T-2 is utilized in this case, and it controls the heating coil control valve only. Transformer XF-1 is energized whenever the unit fan is running, and the transformer's output voltage, or secondary voltage, powers the damper actuators and the control valve actuators. Ventilation control contact TC-1A interrupts the circuit to the damper actuators during unoccupied periods to keep the fresh air damper closed. Since this contact is positioned after the connection for the valve actuators, the valve actuators may continue to operate at night. This ensures that heating coil valve V-1 will open whenever night thermostat NT-1 calls for heat in the wintertime. The firestats and freezestat are the same as before, as is the remote alarm circuit for annunciating firestat or freezestat actuation. Should the firestats or freezestat sense excessive temperatures, the fan will be stopped. Outdoor (fresh air) damper actuator M-1, which is normally closed, spring-returns to its normal position (damper closed) anytime power from control power transformer XF-1 is not available. Return air damper actuator M-2, which is normally open, spring-returns to its normal position (damper open) at the same time.

The foregoing controls for an air-handling unit may be repeated several times throughout the job, either identically or with variations, for additional air handlers serving various parts of the building. For identical units, the same controls sketch can be used but marked to indicate for how many it is typical. Where departure from the typical sequence of operation occurs, a separate sketch is advisable.

PERFORMANCE SPECIFICATION FOR A HEAT EXCHANGER

Other types of equipment are approached in a similar manner. Consider the generation of hot water for use in the heating coil of an air handler. An operational description for a shell and tube steam-to-water heat exchanger—or converter, as it is frequently called—might read as follows:

> Hot water leaving the heat exchanger shall be controlled such that the hot water discharge temperature set point shall be reset inversely with the changes which occur in the outside air temperature. As the outside air temperature rises from 10 to 60°F, hot water leaving the unit shall be reset from 190 to 100°F through modulation of the heat exchanger steam control valve. Continuous readout of outside air and hot water temperatures in degrees Fahrenheit shall be provided on the face of a local control panel. Automatic positive shutoff of the steam control valve shall occur whenever the hot water pump serving the heat exchanger is deenergized. Above 70°F outside air temperature, the hot water pump shall shut down.

Pneumatic Solution

This controls operational description is first approached here pneumatically. As shown in Fig. 2-3, receiver controller RC-1 is piloted by temperature transmitters TT-1 and TT-2, which also provide a continuous linear pneumatic signal to temperature-indicating gauges TIG-1 (hot water) and TIG-2 (outside air), which are mounted in the door of the control panel. The signal from TT-2 also indexes PE-1 for stopping of the hot water pump at 70°F outside air temperature. RC-1 is reverse-acting, and so when the temperature rises above the set point, the receiver controller will remove air from the diaphragm of control valve V-1, allowing the valve to move toward the closed position under the influence of its actuator spring. Solenoid air valve EP-1 exhausts air from the actuator of V-1 anytime the hot water pump is shut down. EP-1 accomplishes this by blocking the signal from RC-1 while completely dumping all the compressed air from the actuator of V-1. During setup, the necessary adjustments are made to commission RC-1 for the rescheduled set points as required by the specification. This description of the heat exchanger control is sketched out, and devices are identified for takeoff and later inclusion on Form 3-1, a sample equipment room labor and materials estimate sheet that appears in Chap. 3.

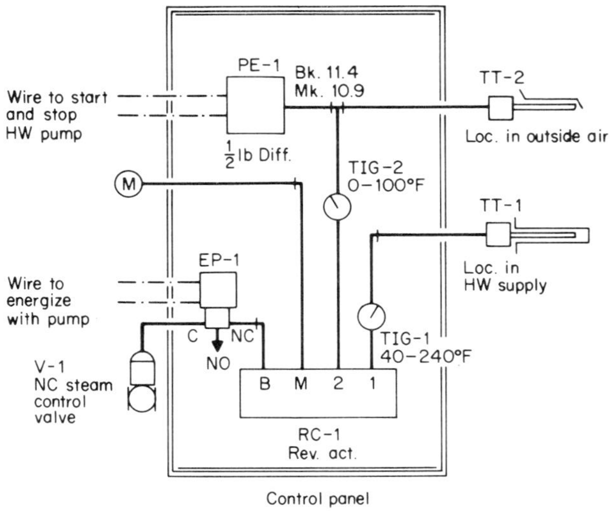

Reset schedule	
OA	HW
10°F	190°F
60°F	100°F

Figure 2-3 Heat exchanger pneumatic controls takeoff sketch.

Electrical Solution

Figure 2-4 shows an electrical approach. Electronic controls are used to obtain a readout and a sliding-scale set point. As can be seen in Fig. 2-4, variable-output electronic temperature sensors provide linear input signal pilot control of TC-1 through their input and display modules ID-1 and ID-2. Low-voltage power is produced by power module PM-1, and logic control and linearized set-point rescheduling are handled by control module CM-1, subject to interlock with the hot water pump motor starter. Output to normally closed control valve V-1 issues from output module OM-1, and start-up and shutdown of the hot water pump are accomplished through the relay contacts in auxiliary module AM-1. All modules are miniaturized and interconnected with each other in a common factory enclosure through plug-in-style terminal strips that emanate from a common motherboard with printed-circuit-style backplane wiring. A temperature display integral with modules ID-1 and ID-2 provides a readout of the temperatures involved, and calibrations and adjustments to set point and set-point shift authority are presented for operator-machine interface on the front of control module CM-1. Temperature controller TC-1 is self-contained. This type of controller can be obtained from a number of manufacturers and is available in various state-of-the-art forms which resemble the one described

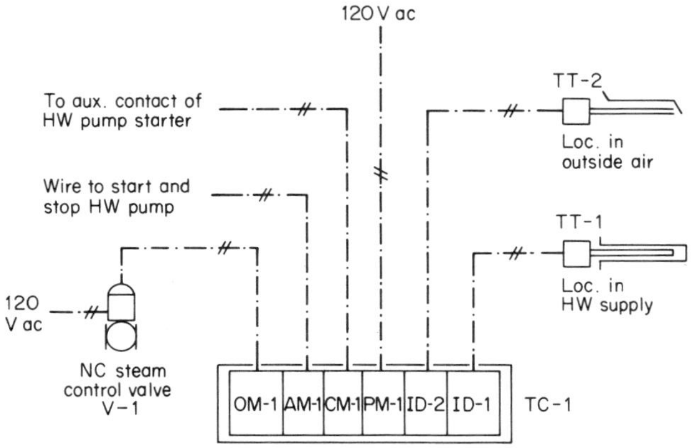

Reset schedule	
OA	HW
10°F	190°F
60°F	100°F

Figure 2-4 Heat exchanger electric controls takeoff sketch.

here. Often, optional function modules can be purchased as individual items. As a result those modules that are not required can be omitted, and blank-off module fronts can be used until more modules are added.

CUSTOMER-INSTALLED DAMPERS AND CONTROL VALVES

Two items in the air handler and heat exchanger control schemes that require sizing for proper estimating are the customer-installed dampers and control valves. Often, controls can be batch-applied for similar functions appearing on various pieces of equipment, but control valves and dampers require sufficient preengineering at bid time to establish the sizes and flow characteristics needed to meet the specific requirements for each application. Using the information from the mechanical contract drawings, make it a practice to transcribe the sizes right onto the control sketches. Costs vary with size, and where a significant number of dampers or valves are involved, it is important to determine how large they need to be.

Dampers

Automatic dampers are normally identified on the mechanical contract drawings. When the dampers are placed within ductwork for which dimensions have been given, you can use the dimensions to help establish damper sizes. If such dimensions are not shown, check the scale of the drawing being referenced, and take the measurements yourself. You will be submitting the sizes that you propose to furnish before the dampers are ever ordered anyway, and the sheet metal contractor will be asked to verify the sizes you have selected.

Two-position control dampers are usually of the parallel-blade design, where all blades rotate in the same direction upon opening or closing. Opposed-blade dampers are generally preferred for modulating action where air flow is to be controlled proportionately. This style of control damper usually exhibits better control characteristics than the parallel-blade design, which can be seen graphically in a plot relating damper opening to air flow. Such modulating control dampers are routinely found on the suction side of the air handler in the mixing plenum that controls fresh air and return air intake to the unit and relief air leaving the building. The controller for the dampers might be a mixed air control, a discharge air control, or a space thermostat when used with appropriate high- and low-limit controllers.

Some added features may be specified also, and these must be taken into account because they add to the cost of automatic dampers. Such items as blade edging and blade end seals make for tighter closeoff and help minimize the infiltration of outside air when the fresh air and relief air dampers are in the closed position. Special framing or blade construction adds to damper cost too, so a thorough understanding of the damper specification is necessary. A need for significant quantities of dampers warrants obtaining confirming quotes from damper manufacturers.

Control Valves

Control valves can vary widely in price. Valve body materials are normally called out in the spec. Customarily, bronze-bodied control valves 2 in or smaller with threaded connections are used for controlling hot water for heating and chilled water for cooling at air handling unit coils. Low-pressure steam for heating is readily handled by these valves also. Iron-bodied control valves 2 1/2 in or larger with flanged connections are also regularly employed. For water service, two-way or three-way patterns can be selected. Two-way straight-through types can be applied where head pressure is not excessive and means of pump pressure relief is provided for. Three-way styles permit constant circulation of water through the valve for coil flow-through or coil bypass and so can be installed in a broad range of applications. If possible, three-way control valves should be used in a mixing arrangement; otherwise, the more expensive diverting configuration will have to be used. When line pressure can be presented to both sides of the control valve's internal plug and seat, an equalizing effect permits use of the simpler and more economical single-plug three-way control valve. This is typical of the mixing piping arrangement shown in Fig. 2-5*a*. The valve's single plug is positioned by the valve actuator between the two internal ports, or openings, of the valve to regulate how much

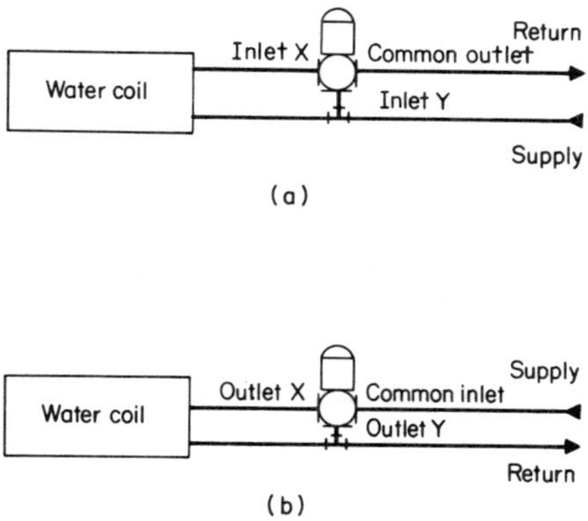

Figure 2-5 Three-way control valves. (*a*) Single-plug valve; (*b*) diverting-type valve.

water bypasses the coil and how much is allowed to pass through. Note that the supply pressure approaches the valve toward openings X and Y. That which enters the valve at X has been through the coil, but is approximately the same pressure as that at opening Y, except for a pressure drop through the coil. However, in Fig. 2-5*b*, the supply pressure arrives at the common port or opening of the control valve, flows through the valve first, and is then diverted to either X or Y. In this case a diverting-type three-way control valve with double plug and seat is necessary. With this type of valve, the incoming pressure is impressed against opposing plugs at openings X and Y to create balanced forces within the valve and ensure better control of pressures which the actuator must overcome.

Sizing of the control valve is dependent upon pressure and flow. Pressure is not a problem since in commercial heating and air conditioning most 150-lb-rated control valves will suffice. This leaves flow as the item to be reckoned with. Control valves are rated with a C_v number, which represents the valve's capacity to pass 60°F water with a pressure drop of 1 lb of pressure at the quantity of gallons per minute represented by the C_v number. The C_v number simplifies control valve selection since selection may be made on the basis of the valve's C_v number. This number, of course, applies to the valve when it is in the full-open position. To see how any given valve will accommodate various flow requirements at differing pressure drops, a formula is used, with entry data supplied by the contract drawings or project specifications. This formula aids in the selection of the correct valve to handle the flow requirements of the coil for which it is intended. For control valves handling water flow, the equation reads:

$$C_v = \frac{Q_w}{\sqrt{P_d}}$$

where Q_w is the water flow in gallons per minute which the valve in its full-open position must allow to flow through and P_d is the pressure drop that can be tolerated through the valve. The equation states that to find the C_v requirement for the proper valve, the water flow in gallons per minute must be divided by the square root of the allowable pressure drop. For example, if a coil in an air-handling unit needed a maximum of 56 gal/min with no more than a 2-lb pressure drop through the valve, the C_v requirement would be found by dividing 56 by 1.4, which equals 40. The control valve to select would be the one whose C_v is closest to 40 without going below it.

For control valves handling steam, the formula looks like this:

$$C_v = \frac{Q_s\sqrt{V}}{63.5\sqrt{P_d}}$$

where Q_s is the steam in pounds per hour that the control valve must be able to pass in the full-open position, V is the specific volume of the steam at the pressure when it reaches the valve, and P_d is the allowable pressure drop through the valve. Some similarities between the two equations for water and steam are obvious, but the one for steam is a little more involved, though by no means difficult. With a requirement for 250 lb/h of 5-lb steam with a maximum 2-lb pressure drop through the valve, the solution to the equation would read 250 multiplied by 4.518 divided by the product of 63.5 multiplied by 1.4, or 1129 divided by 89. This reduces to 12.7, which represents the smallest acceptable C_v.

Coil flow capacities in terms of gallons per minute or pounds per hour of steam are determined from the contract drawings or from coil data pertinent to the specifications for the air-handling unit. In the case of a steam coil, sometimes the Btu is the unit of measure given; Btu stands for British thermal unit and is an expression of the coil's heating capacity in actual heat output. Dividing the Btu's by 1000 yields a number which can be considered the equivalent of the required pounds of steam per hour for steam valve sizing.

The square roots of specific volumes of steam are shown in Table 2-1, which lists the properties of saturated steam for gauge pressures through 20 lb. Once the C_v number is known, manufacturers' valve data sheets can be consulted to select a control valve with a C_v number large enough to suit the needs of the application at hand.

PERFORMANCE SPECIFICATION FOR A CENTRIFUGAL WATER CHILLER

The specification that follows details the sequence of operation pertaining to chilled water and condenser water control. Not only is temperature control spelled out, but electrical interlock is covered too, so that functionally associated pieces of equipment will operate at the same time. For example, it would certainly be unnecessary for the cooling tower fan to come on unless the condenser water pump was working. And it would be equally undesirable for

Table 2-1 • Properties of Saturated Steam

Gauge Pressure, lb in^2	Steam Temp., °F	Specific Volume, V	$\sqrt{V}$
1	215.3	25.2	5.02
2	218.5	23.78	4.876
3	221.5	22.57	4.75
4	224.4	21.4	4.626
5	227.1	20.41	4.518
6	229.8	19.45	4.41
7	232.3	18.64	4.317
8	234.8	17.85	4.225
9	237.1	17.16	4.142
10	239.4	16.49	4.061
11	241.6	15.9	3.987
12	243.7	15.35	3.918
15	249.8	13.87	3.724
20	258.8	12.0	3.464

the chiller to operate without the chilled water pump. A sample performance narrative involving a centrifugal water chiller might read as follows:

> Start-up of mechanical cooling shall occur automatically on a rise in the outside air temperature to 60°F. At such time, the water chiller shall be indexed to begin sequential start-up of the evaporator and condenser water circulators. When flow has been confirmed at the machine by differential-type evaporator and condenser water flow proving devices, the chiller shall be enabled to run, and operation shall begin, provided all internal safeties are satisfied. Once the chiller has started, chiller capacity shall be controlled by regulating the temperature of the leaving chilled water through modulation of the refrigerant inlet vanes. The cooling tower fan operation shall be interlocked with the condenser water flow, and the tower fan shall cycle on as required to maintain a temperature of 90°F for water returning to the machine, provided the tower bypass control valve has been first positioned for full flow over the tower. When the outside air temperature falls below 60°F, all operation of mechanical cooling equipment shall cease.

Pneumatic Solution

Figure 2-6 shows pneumatic controls that have been applied to meet this specified sequence of operation. Thermostat T-1, which senses outside air temperature, will "make" PE-1 at 60°F on a rise in temperature. This closes a starting cycle circuit within the chiller control panel. An internal contact closure is initiated by the starting cycle to terminals 3 and 4 of the chiller panel so that operation of the chilled water pump may begin. This

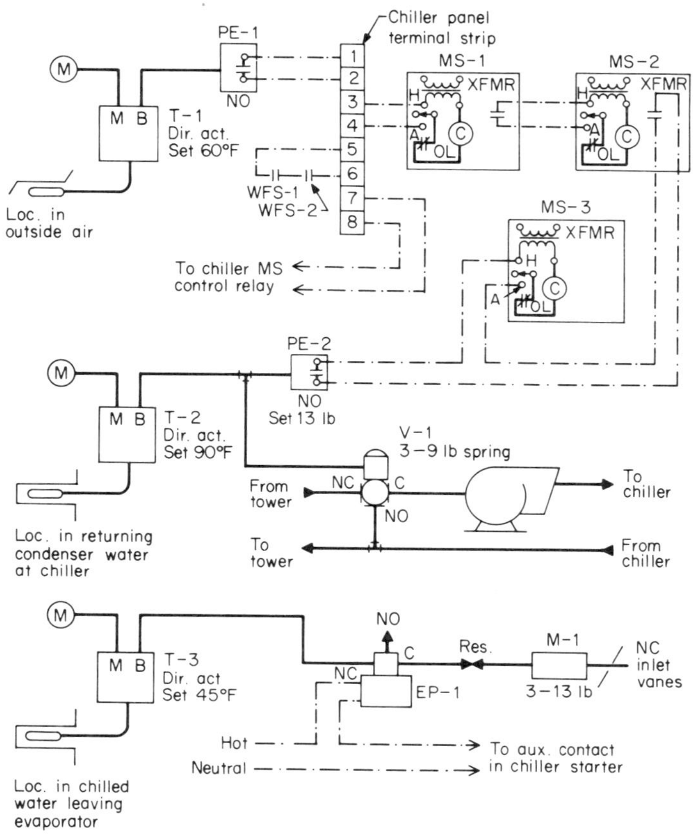

Figure 2-6 Chilled and condenser water controls.

is accomplished by energizing motor starter MS-1. An auxiliary contact in MS-1 closes simultaneously to institute operation of the condenser water pump through condenser water pump motor starter MS-2, and an auxiliary contact in MS-2 closes to allow energization of cooling tower fan starter MS-3. Now the chiller waits momentarily for proof of flow through the closing of water flow switches WFS-1 and WFS-2 in the chilled and condenser water lines, respectively. Once that happens, an internal contact in the chiller panel closes to complete a circuit to terminals 7 and 8, and the remote chiller motor starter is allowed to energize. At the same time, EP-1 is picked up and places the chiller capacity inlet vanes' actuator under the control of thermostat T-3. As the condenser water temperature rises, condenser water thermostat T-2 modulates tower valve V-1 for tower flow, and PE-2 "makes" to start the cooling tower fan once V-1 is fully opened to flow to the tower. When the outside air temperature falls below 60°F, T-1 "breaks" PE-1 and operation of all equipment ceases. EP-1 bleeds down the signal line to the vane actuator, ensuring that damper actuator M-1 will be at its minimum position the next time the cooling equipment must restart.

Chiller terminal strip numbering varies widely, and so reference to chiller manufacturers' wiring diagrams is essential at the time that engineering of the job occurs. Also, MS-1, MS-2, and MS-3 are usually furnished by others, and T-3, EP-1, and M-1 quite often come with the chiller. Additionally, WFS-1 and WFS-2 are frequently part of the trim furnished by the chiller manufacturer. But diagramming a scheme such as this permits visualization of the controls involved and facilitates coordination with other equipment suppliers before bid day. Ordinarily, the equipment room labor and materials estimate sheet shows T-1, PE-1, T-2, PE-2, and V-1, assuming that the balance of control devices will be supplied by others. Pneumatic tubing will need to be run for all pneumatic devices, even T-3, EP-1, and M-1, because T-3 is often shipped loose for mounting downstream in the chilled water line by the controls contractor. And so, although T-3 is not furnished by the controls contractor, the controls contractor needs to estimate labor for its installation. The two thermal wells for the pipeline thermostats, one furnished by the controls contractor, are installed in the pipeline by the mechanical piping contractor. V-1 will most likely be a sizable three-way valve, and so gallons per minute (gpm) requirements must be checked and the valve sized accordingly. For condenser water service such as this, a butterfly-type three-way valve can often be used, but the specs must be checked to make sure.

Wiring consists of short runs of flexible conduit for PE-1 and PE-2, for these relays will have to be located within 5 ft of the devices they control. WFS-1 and WFS-2 will probably be mounted and pressure-tubed by the chiller manufacturer or the mechanical contractor, but may not be wired, and so verification of this point is in order. If motor starters MS-1, MS-2, and MS-3 are all within a motor control center, internal interlocking will be all that is necessary, except for the run to the chiller in EMT and the run to the remote chiller starter. The hours for both pneumatic and electrical labor should be entered on Form 3-1, an equipment room labor and materials estimate sheet appearing in Chap. 3, if the control wiring is part of the control work. Hours can be selected from Table 3-1, a chart listing equipment room labor-hours that appears in Chap. 3, and entered in the appropriate columns on Form 3-1. Labor should be entered for both tube and wire for pneumatic and electric devices for which the controls contractor has complete installation responsibility. Labor *only* should be entered for devices provided by others that the controls contractor must connect, mount, or both.

Electrical Solution

Electric controls could easily be substituted for pneumatic controls to accomplish the same functional response. T-1 might be an electric remote-bulb thermostat whose internal contacts make the circuit back to terminal strip numbers 1 and

2. Conduit and wire could replace the pneumatic line between T-1 and PE-1. Condenser water control valve V-1 could be an electric modulating assembly with a limit switch to start the cooling tower fan. In such a case, T-2 would be an electric proportional thermostat. And an electric modulating actuator might be required in lieu of pneumatic M-1. Through appropriate interlocking, an electric damper actuator could be made to drive to the minimum position whenever the chiller was shut down. Estimating would proceed in an all-electric vein, and the job could be supervised rather than installed, where the controls contractor engineers the system, provides the controls, and supervises the installation by the jobsite electrical contractor. The controls portion of the specification details whether the control system is to be installed by the controls contractor or by the jobsite electrical contractor. When a system is supervised, administrative duties, engineering, supervision, and checkout time need to be included in the estimate.

Some comments about supervised installations are appropriate here, because in estimating supervised systems there are certain functions related to the installation of controls that remain with the controls contractor. When the controls contractor quotes a supervised installation, the estimate usually includes labor-hours for working on electric systems, even though another contractor's electricians will be mounting and wiring the devices. As control experts, controls contractors are quite often expected to (1) link dampers to their electric actuators and (2) calibrate and commission the system. Generally, controls contractors include labor-hours for jobsite supervision to answer questions as the installation progresses, plus time for linking dampers and setting up and adjusting the job. Most sheet metal and electrical installers shy away from damper linkup because of inexperience in this area. As controls estimators, controls contractors need to appreciate the difficulty involved and the expertise required.

Damper Linkup to Electric Actuators Damper actuators require a certain amount of attention for adjustment of linkages and verification of proper opening and closing in response to controller signals. In the case of an electric actuator, it is best to apply electric power and observe how the arc of the actuator crank arm relates to the desired movement of the damper crank arm. At the time the linkages are connected, the controls contractor has certain options available for driving the damper to be controlled. Consider a simple two-position fresh air damper with a drive shaft extended through the side of the duct for connection of the actuator. The blades of the damper rotate 90° from the full open to the full closed position, as does the extended drive shaft. Many of the popular models of actuators deliver a 160° rotation in the process of moving from one end of their stroke to the other. The linking-up process thus involves proper placement of connecting ball joints on actuator and damper shaft crank arms to allow full actuator drive shaft rotation to deliver no more than a 90° arc at the damper shaft axis. Additionally, it must be decided whether to use the same or opposite clockwise or counterclockwise pivoting to facilitate the normally open or normally closed damper position.

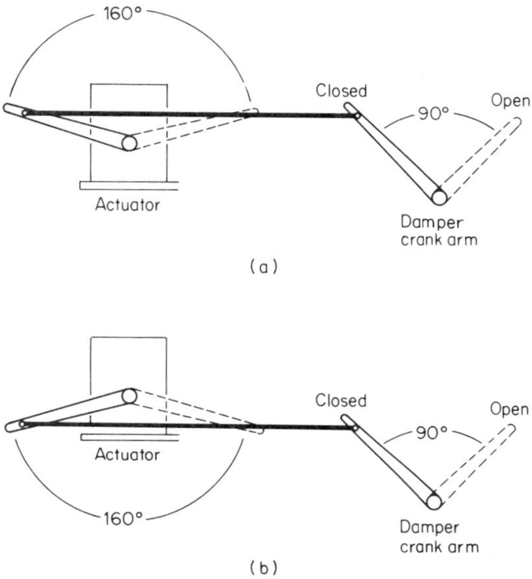

Figure 2-7 Actuator crank arm positions. (*a*) Overhand arc; (*b*) underhand arc.

The sketches in Figs. 2-7 and 2-8 illustrate actuator crank arm positions and damper crank arm position and are intended to guide the control technician through damper setup and checkout and to lead to a better understanding of the labor-hour allowances made for performing a thorough damper actuator checkout. A gear-driven electric actuator is the type shown in Figs. 2-7 and 2-8.

Actuator crank arm positions In Fig. 2-7, note the flexibility obtained through placement of the actuator crank arm for either the overhand (Fig. 2-7*a*) or underhand (Fig. 2-7*b*) arc to compensate for either clockwise or counterclockwise rotation of the actuator drive shaft. Assuming this is a spring-return two-position damper actuator, the deenergized state positions the damper crank arm for a closed damper, and when energized, the actuator cranks to move the damper toward the open position. The relatively longer length of the damper crank arm permits placement of the ball joint rod connector out toward the end so that greater travel is required to achieve the full 90° movement at the damper shaft. Conversely, by moving the connecting ball joint at the actuator more toward the axis end of the crank arm, the travel of the pushrod is effectively reduced, allowing the 160° rotation of the actuator shaft axis to be mated to the 90° rotation of the damper shaft axis. This gives almost a 2 to 1 stroke ratio between the actuator and damper, which results in a strong power stroke for overcoming any hysteresis that might be present in the bearings or side seals of the damper. The spring-return actuator closes

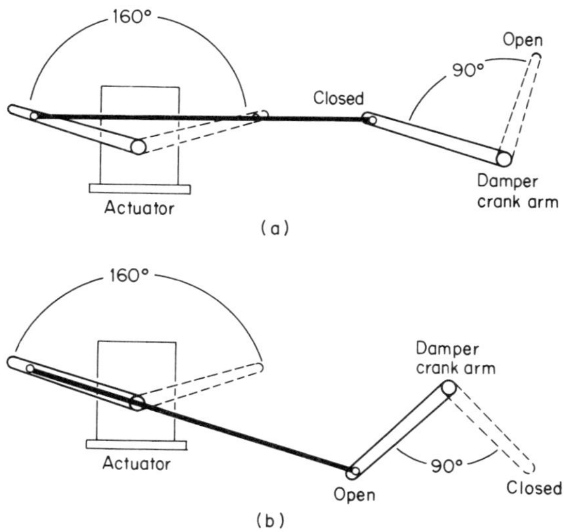

Figure 2-8 Damper crank arm positions. (*a*) Normally closed operation; (*b*) normally open operation.

the damper upon loss of electric signal and cranks to open the damper upon restoration of its power circuit. Ideally, the linked damper should be stroked from one extreme to the other so that it can be observed whether there is an even stroke without binding or jamming. When the electric wiring incidental to control of the actuator has been completed, the cover plate can be opened and one wire lifted from its terminal repeatedly to simulate actual damper operation so that adjustments can be made. If the fan in the air-handling unit is running, it may be advisable to open the return air plenum doors until all dampers are properly adjusted to prevent the creation of a vacuum in the return air plenum which might damage the ductwork. Frequently, the technician must conduct the setup work before the electric wiring is completed or during a period when the air-handling unit fan cannot be run. For this purpose, a 50-ft extension cord of 14-gauge wire suitable for rough service can be attached to a 24-V transformer through a 4-in. electric box with an on-off toggle switch. This is an ideal portable low-voltage power source for powering the actuator for test purposes. For line voltage models, a second set of test leads, connected in parallel to the primary side of the transformer within the outlet box and controlled by a second switch, can be used in the same fashion.

Damper crank arm positions The same overhand or underhand rotational arc can also be adapted to the crank arm for the damper shaft. This allows for setting up the damper for normally closed or normally open operation, as with an outside air or a return air damper. In an economizing sequence of operation, the return air damper assumes a partially closed position in response to the fresh air damper assuming a partially open position. The dampers move in unison so that the air being let in through one replaces the air being blocked out by the other. In Fig. 2-8*a*, the damper is linked for normally closed operation, as would be the case with an outside air intake damper. However, in Fig. 2-8*b*, the damper is linked for normally open operation, as would be the case with a return air damper. Now identical actuator shaft rotations deliver different positions of their respective dampers. The interconnecting pushrod pivots at either end in a ball joint connected to the crank arm.

Pneumatic damper actuators, too, need to be stroked for proof of proper linking. By and large, the piston type of pneumatic actuator is most commonly used in the controls industry for commercial applications. A swivel joint on the end of its pushrod permits pivoting of the drive pin, or swivel mounting of the actuator allows it to reposition itself as the crank arm rotates. In either case, compressed air should be applied and the motor stroked to make sure that the damper moves freely from one extreme to the other. When compressed air is not available, a pneumatic squeeze bulb is used to pump up the actuator so that coupling and alignment of the interconnecting linkage can be observed. Unlike the electric actuator, the pneumatic actuator has no crankshaft to create an arc of travel for the actuator crank arm, and adjustment is probably easier. The pneumatic actuator used for control of dampers takes advantage of the buildup of compressed air behind a diaphragm attached to a pushrod. Thus, straight-forward motion is imparted.

Setting Up and Adjusting Actuators In some cases, actuator mounting and linking templates are shipped with the damper motors, but there is no substitute for physical operation of the actuator as it is attached to the damper drive shaft. Only in this way can you be sure that the dampers are properly linked and will work smoothly when the system is put into operation. And once you have set them up using true motive force to observe accurate damper performance, you won't have to go back when the system fan is running to verify that the linkage adjustments are correct.

When considerable damper work is involved, the controls estimator who is familiar with the intricacies of damper setup is more likely to allot a proper number of labor-hours for this work than the estimator who has less experience. And in estimating hours, an item not to be overlooked is the height at which damper linkup is to take place. Suffice it to say, linkup is much simpler when the air handler is at floor level and the work can be walked up to. More time is required if the unit is hung from the overhead structure of the building and the work has to be accomplished from a lift. Timing is a factor too, as owner occupancy of the floor space below can cause difficulty in getting lifts into position. It is important to make sure that the time frame allowed for in the estimate is carried over into the job schedule so that high work is done when the floor space below is open. Consideration, too, must be given to the installation and setup of damper actuators while the unit is still on the floor and before it is lifted into its position high above. The portable extension cord power supply serves particularly well here, permitting the technician to finish the work at floor level. The well-planned estimate is effective only when it becomes part of the plan of attack once the bid is won. Those special laborsaving ideas envisioned at the time the estimate was prepared should be clearly noted on the estimate package for implementation by the construction management team to ensure that installation performance meets the objectives of the estimator.

3

Estimating Equipment Room Temperature Controls

It is important to determine the devices that need to be mounted and connected on equipment for which an in-room footage allowance of tubing or wire and/or conduit may be assigned. Later, this information will be helpful when entering the costs of devices, installation material, and labor on Form 14-1, a sample bid summary sheet that appears in Chap. 14. Form 3-1, a sample equipment room labor and materials estimate sheet, can be used for recording information about these devices and is designed to list on one page the costs associated with fan room or equipment room work, whether the control devices are electric or pneumatic, or both.

THE EQUIPMENT ROOM LABOR AND MATERIALS ESTIMATE SHEET

At the top of Form 3-1, there is a place to enter the job name, the date of the estimate, the estimator's name, the page number, the room or unit number, and the floor or area. In the column headed "Control," all the model numbers of control devices which must be furnished to accomplish the work in this particular area are written in. In the two columns headed "Pneu." and "Elect." the estimator specifies whether the device is pneumatic or electric. The quantity required for the area under consideration is listed in the column headed "Qty." The cost per control in dollars is entered in the column headed "Cost/ea.," and the total cost, which is arrived at by multiplying the quantity by the cost per control, is entered in the column headed "Total cost." In the next column, headed "Labor/ea.," the number of labor-hours required for the control is entered. The allowable hours can be found from Table 3-1, a chart listing equipment room labor-hours that appears later in this chapter. The allowable labor-hours multiplied by the quantity gives the total number of hours, which is entered in the next column, headed "Total labor."

Because different types of controls, both pneumatic and electric, are likely to be used on any one job and because various methods of installing tubes

Equipment Room Labor and Materials Estimate Sheet

Job name ____________________ Date __________

Estimator ____________________ Page _____ of _____

Room or unit number ____________________ Floor or area __________

Control	Pneu.	Elect.	Qty.	Cost/ea., $	Total cost, $	Labor/ea., h	Total labor, h	A	B	C	D	E
Panel rings	—	—				One		—	—	—	—	—
Total				//////		//////						

Transfer total equip. room control device costs $ ________ to bid summary sheet, line 2

Transfer total equip. room hours _____ (X 1.3)* = ________ to bid summary sheet line 11

(A) Elect. controls wired in EMT:	_____ X 30 ft/ea. =	_____ ft	EMT and wire
_____ ft EMT and wire (X 1.3)* =	_____ ft X $	_____ /100 ft =	$ _____
(B) Elect. controls wired rigid:	_____ X 30 ft/ea. =	_____ ft	rigid and wire
_____ ft rigid and wire (X 1.3)* =	_____ ft X $	_____ /100 ft =	$ _____
(C) Pneu. controls piped in EMT:	_____ X 30 ft/ea. =	_____ ft	EMT and tubes
_____ ft EMT & tubes (X 1.3)* =	_____ ft X $	_____ /100 ft =	$ _____
(D) Pneu. controls tube bundled:	_____ X 30 ft/ea. =	_____ ft	tube bundled
_____ ft tube bundled (X 1.3)* =	_____ ft X $	_____ /100 ft =	$ _____
(E) Pneu. controls tube copper:	_____ X 30 ft/ea. =	_____ ft	tube copper
_____ ft tube copper (X 1.3)* =	_____ ft X $	_____ /100 ft =	$ _____

Transfer total installation material costs to bid summary sheet line 7 = $ _____

* Use factor only if area over 750 ft^2

Form 3-1 Equipment room labor and materials estimate sheet.

or wiring are desirable, five columns, headed by the letters A through E, are provided for citing the installation materials associated with the controls on the sheet. The same number that appears in the quantity column should also be entered in the appropriate lettered column, so that when these last five columns are totaled, the respective quantities can be assigned to the proper places in the lower portion of Form 3-1 for pricing. On the form, column A represents EMT and wire and column B represents rigid conduit and wire; one of the

Table 3-1 • Equipment Room Labor-Hours for Control Devices

Control	EMT Elect.	Rigid Elect.	EMT Pneu.	Bundled Pneu.	Copper Pneu.
Sensor or thermostat					
Remote-bulb	4.0	5.0	4.0	2.5	3.5
Rigid-stem	2.5	4.0	3.5	2.0	3.0
Freezestat	4.0	5.0	—	—	—
Firestat	2.5	4.0	—	—	—
Enthalpy	4.5	5.0	5.0	4.0	6.0
Duct smoke detector to starter coil	4.0	5.5	—	—	—
Add for 120-V feed	2.5	4.0	—	—	—
Add for FACP tie-in	2.5	4.0	—	—	—
Damper motor	4.0	5.0	4.0	2.5	4.0
Control valve	5.0	6.5	4.0	2.5	4.0
PE and EP*	1.0	1.0	2.5	2.0	4.0
Water or air flow switch					
Paddle	2.5	4.0	—	—	—
Delta-P (ΔP)*	2.5	4.0	1.5	1.0	1.0
Static sensor	4.0	5.0	4.0	4.0	5.0
Pressure control	5.0	6.5	5.0	4.0	5.0
Receiver controller and enthalpy controller	3.5	4.0	4.5	4.0	4.5
Relay	3.5	5.0	4.0	2.5	4.0
Switch	2.5	4.0	2.5	2.0	4.0
Time clock*	4.0	5.0	4.0	1.0	1.0

* Add the labor-hours for the electrical and pneumatic entries together if the device requires both.
Note: The allowances include time for mounting, running 30 ft of interconnecting tube or wire, calibrating, and setting up.

two will be applicable to any device requiring wiring. Column C represents EMT and tubing, column D is for bundled tubing, and column E represents copper tubing; these three are associated with pneumatic controls. While the spec may allow pneumatic tubing in EMT, there may be cases where it is preferable to run just a single copper line, or maybe two, instead of the EMT and polyethylene tubing. From Table 3-1, which lists the equipment room labor hours for control devices and appears above, it becomes evident that the one or two lines required for remote-bulb or rigid-stem equipment room sensors or thermostats can be run more economically with two bare copper lines than with EMT and polyethylene. Thus, Form 3-1 provides the flexibility to estimate either or both and to make the cost of these items a part of the total installation materials costs for the equipment room under consideration. This greatly simplifies estimating controls for equipment rooms, and allows the same sheet to serve as a pneumatic estimate or an electrical estimate or both.

After the respective columns are totaled, the totals are entered in the lower portion of Form 3-1, where there are reminders to transfer the dollars or hours, as the case may be, to the appropriate line on Form 14-1, a bid summary sheet. Additionally, in the lower part of Form 3-1, labor-hours and installation materials costs are adjusted upward whenever equipment rooms in excess of 750 ft^2 are involved. Factoring with a multiplier of 1.3, *only if* the equipment room or area of a large mechanical room floor is over 750 ft^2 in size, allows for the increased amount of installation materials likely to be required and the additional labor-hours associated with this increase. As respective quantities of EMT, rigid conduit, wire, and bundled or copper tubing are brought to the right-hand side of the estimate sheet for totaling, a ready reference is created that shows how much and what kinds of installation materials were figured.

To make sure that you take into account the labor-hour and installation material costs associated with devices furnished by others, you can also list such devices on Form 3-1 and merely omit the costs of the control devices. Such a procedure ensures that all the work required of you within any given area is listed on one sheet, regardless of who must furnish it.

You should use a separate equipment room labor and materials estimate sheet for each equipment room in a building and for each major piece of equipment which, because of its size and the area of square footage it inhabits, would require separate figuring. These sheets provide a quick reference to segmented listings of work and materials for each work area or piece of equipment. Such an approach makes cost validation of work performed much easier, particularly if installers are instructed to code their job time to specific segments that match the estimate sheets. This can be helpful in determining why jobs run over in cost. Good estimating should incorporate feedback for cost comparison. Such an ongoing improvement-oriented method produces quality estimates. In addition, assigning labor-hour goals to installers allows each person to compete against the estimate in an effort to reduce labor-hours, improve performance, and gain recognition.

It is recommended that the complete model number of the control device be entered in the column headed "Control" on Form 3-1. Since in order to arrive at the complete model number some details of engineering have to be considered, you can be sure that what you've selected through labeling of symbols on the takeoff sketch will perform as required and meet the specification.

Unit Allowance

Using unit allowances of labor-hours for equipment room devices, and for materials as well, forces the controls contractor to overstate in some cases and understate in others. But the inclusion of labor as well as raceway and tube or wire in the estimate for unit work proves to be remarkably accurate, since scaling becomes exceedingly difficult because of the complexity of fan room installations and the angles to be expected when running conduit or piping on ducts, across units, and along wall surfaces where offsets regularly occur. Determining total footage requirements is impractical in such a situation, and using unit allowances is more practical and feasible. It results in faster estimating with satisfactory accuracy. The labor-hour and materials allowances apply to work within the equipment room only, as the 30-ft distance does not include tubing or wiring outside of the equipment room.

Segmenting

It has been noted that the 1.3 factor is used for equipment rooms over 750 ft^2. For those equipment rooms over 1200 ft^2, the area should be segmented and the segments considered as separate equipment spaces. Large mechanical-level floor areas in high-rise buildings can be more systematically estimated when segmented into smaller areas, which can then be approached as individual equipment areas. Additionally, the segmenting approach to extremely large equipment areas helps to locate the work areas with respect to the equipment to be found in those areas, making future ordering and labeling of packages of materials easier.

Panel Rings

In many equipment rooms a control cabinet, or control panel, in which non-unit-mounted devices can be grouped is used. The EMT and tubing and/or wire allowances for the control room devices already provide for routing of the tubes or wires to other controllers within the equipment room, and many of them could be in the control panel. While estimating the control panel is approached separately, the cost and installation of the enclosure for the panel are entered on Form 3-1 because the enclosure almost always is installed ahead of the interior of the panel, which will have the panel-mounted devices on it. And because the most efficient kinds of control room enclosure-type panels are installed in two stages, the need to allow for the cost and installation of the enclosure on Form 3-1 is quite practical. Remember, during rough-in of the equipment room, running the piping or EMT to the panel necessitates punching or drilling holes for tubing and raceway entry into the box. A separate ring installed early in the rough-in portion of the work provides the installers with a target to work toward with their pipes and conduits, and facilitates placement of the subpanel within the ring after the wires and tubes have been pulled in. Thus at the bottom of the "Control" column of Form 3-1 are the preprinted words "Panel rings." An hour for the placement and mounting of each ring is allotted.

Mounting

Fan room work is generally a significant portion of the labor estimate for temperature control work because so many of the control devices must be mounted on the air handler itself. The controls companies will mount the control equipment they furnish, but perhaps not all of it. Consider devices such as smoke detectors, firestats, air flow switches, and remote-bulb thermostats. Unless the responsibilities for mounting are clearly delineated on the quotation, the controls contractor may very well inherit the job of mounting these devices in a pneumatic systems installation, even though the wiring incidental to the pneumatic control system is specified to be done by the jobsite electrical contractor. The four devices just mentioned do not utilize compressed air and so do not require a pneumatic connection. They are purely electric devices, except for possibly the air flow switches, which, if of the differential type, require minimal tubing to tap into the duct of the air handler so they can sense the necessary air pressure within the duct.

Thus, on the quotation and at the time of bid, it is wise to announce your position with respect to mounting equipment of this nature. Without stating your position, your bid might seem unnecessarily high if you include mounting and your competition does not. You might lose the bid to a competitor who lets the issue ride, to be resolved later during the installation. You might never know why you didn't get the bid. However, the competitor who got the bid may well have to mount *all* the equipment.

Devices which are usually mounted by the mechanical contractor include valve bodies and dampers. Further, it is customary and appropriate to exclude this work from the equipment room labor and materials estimate sheet to ensure your position. Dampers and valve bodies generally are shown on mechanical plans, and so these are items for which the mechanical contractor will allow labor anyway. Few other control devices will be shown on the contract plans except for room thermostats and possibly static pressure controls and remote-bulb thermostats.

On Form 3-1, there is a place to enter a labor unit beside each control device. If you don't assign labor units to firestats, smoke detectors, remote-bulb thermostats, or air flow switches, you need to note this on the quotation for consistency throughout the estimate and bid. On the other hand, if you do decide to mount this equipment, by indicating this on the quotation, you could well have the edge on someone who does not do mounting. In either event, have your proposal match the estimate. It will save you money and will work to your advantage to have all the details fully itemized. Your options during negotiation can be enhanced by clearly defined work spelled out in a well-prepared confirming proposal. Also, by paying careful attention to the scope of your submission, you can be assured that your customers will understand exactly what they can expect for the price offered.

Pricing

Prices change, and costs rise. Therefore, when presenting a proposal for temperature control work—or for any work, for that matter—it is important to state on the proposal the length of time that the price can remain firm for acceptance. For example, if the length of time is 3 months, a price quoted 6 months ago cannot be arbitrarily accepted today. By specifying a time limit, you can avoid being put in a position of performing work at perhaps an increased cost. A period of 90 days is thought by many to be a fair time for guaranteeing a firm acceptance price. Beyond that, such acceptance is at the option of the offeror.

ASSIGNING LABOR-HOURS

Work done in connection with fan or equipment room devices is generally performed after space rough-in has been accomplished and air-handling or power-generation equipment has been set, ducted, and piped. In the typical equipment room which is ready for control work to begin, raceways have already been stubbed in from adjoining ceiling spaces and can be observed as tubes or wiring which protrudes into the work area and which now needs to be connected and extended to devices or control panels. At this point in the installation, labor is deployed on the basis of the equipment room labor and materials estimate, which was developed before bid day, using unit labor-hours selected from estimating guidelines such as those presented in Table 3-1.

Table 3-1 suggests labor-hours for control devices which are regularly used in an equipment room. For most controls contractors, the labor-hour values are probably very similar, varying only in relation to particular instruments which are known to present special problems in installation and setup. I feel the labor-hours listed in Table 3-1 should be generally applicable to the control devices manufactured by the major companies and routinely found in commercial temperature control applications. As with any guideline, the figures should be modified if necessary to more closely reflect your own experience and individual techniques and work-force management.

These labor-hour values allow for the following:

1. Mounting the control, exclusive of thermal wells and pressure taps for pipeline devices
2. Extending 30 ft of two tubes or two wires in 3/4-in EMT and connecting them
3. Commissioning the instrument, which includes calibration and setup

Thermal wells for pipeline installation and pressure taps for pressure-sensing instruments are usually routinely mounted by the mechanical contractor, as the valve bodies and dampers are, but be sure to clarify this point for your own protection and as a matter of information for those to whom you submit a bid.

Damper Motors and Control Valves

The number of hours allowed for damper motors indicates that the actual in-duct mounting of the damper is to be done by a different contractor, such as the sheet metal contractor; the time shown allows for only mounting, linking, and tubing or wiring the motor.

The same is true for control valves. The actual in-line installation of the control valve bodies is performed by the mechanical contractor, and the hours stated in Table 3-1 allow only for mounting, linking, and wiring or tubing the valve motor, or actuator. The control valves *with* their respective motors should be listed on Form 3-1, but control dampers should be written up separately, identified by where the dampers of various sizes go, and entered as a separate group price on a bid summary sheet (see Chap. 14, Form 14-1).

Tubes and Wires

Shorter runs of greater numbers of tubes and wires are generally encountered within the equipment area where the work is to be accomplished, but systematically allowing for 30 ft of two tubes or two wires in 3/4-in EMT, bundled two-tube polyethylene, or two tubes of hard copper will produce a labor and materials cost for work in the equipment room which comes surprisingly close to the cost of the varying sizes of raceway and tube or wire lengths subsequently purchased to do the job. Only develop a definitive bill of materials for purchasing items for equipment room work if you become the successful bidder and win the contract for the work. For electrical work within equipment rooms, check the project specifications and local codes to be sure rigid conduit is not required. If it is, make sure it is put on your estimate sheet that way. These labor-hour and distance values apply to the equipment room devices that are field-mounted on piping or duct or wall space within the room, not to the devices in the control panels. The values do, however, allow for the lines to get back to the control panel, but panel fabrication, mounting, and connecting will be examined separately.

Thermostats and Sensors

In Table 3-1, the first column lists the controls. Hours are selected by finding the control being estimated and then reading across the table until you come to the column that reflects the installation method required. In the tables, thermostats and sensors which feature a remote bulb with a capillary are grouped together. The hours assigned for this "family" of devices reflect the small amount of extra time needed to string the capillary and set the sensing bulb. Rigid-stem thermostats and sensors need less time because they are self-contained. Freezestats and firestats are purely electric devices, and so no time is listed for them in the pneumatic columns. The same is true of smoke detectors. Notice that extra time has to be added if a remote feed is necessary for the detector and if the detector is to be tied in with the building fire alarm control panel (FACP). The time shown for damper motors allows for linking the motor to the damper too. Control valve bodies are installed into the pipeline by the mechanical contractor, and so the allowances include time for control valve actuator setting and adjusting, but *not* for piping of the valve. Electro-pneumatic relays (EPs) and pneumatic-electric relays (PEs) require both electrical and pneumatic attention, and so the two entries have to be added to get the total time.

Flow Switches

Although the differential-type (ΔP) flow switch is considered to be all-electric, the pneumatic installation by a fitter of the tubes for sensing duct or liquid pressures must be added in as pneumatic time.

Static Sensors and Pressure Controls

The static sensor refers to duct static pressure sensors or controllers mounted on walls or ducts within 6 ft of the point where sensing is required. The hours allowed for vapor or liquid pressure controls include time for setting the device, but installation of pipeline-threaded pressure taps is done by the mechanical contractor. The hours allowed for pneumatic receiver controllers are average times for both one-input and multiple-input models. The times for relays and switches are averages too, insofar as the number of terminals is concerned. The time estimated is for six terminations (tubes or wires), but two or three tubes or wires more or less should not result in any significant deviation.

For controls that use both wiring and tubing, hours for electrical work are omitted in cases where the estimate is purely for pneumatic work and all the electrical work is to be done by others.

Transferring Labor-Hours to Form 3-1

All the hours can be transferred directly from Table 3-1 to the column of Form 3-1 headed "Labor/ea.," and the total labor-hours can be calculated by multiplying this value by the quantity of devices required. Later on in the estimating process, pricing the labor-hours is provided for on the bid summary sheet.

Making Realistic Labor-Hour Allowances

If the unit labor-hour allowances given in Table 3-1 seem high at first glance, remember that the average productive time of an installer during the day is about 6 h. You must take into consideration travel to and from the job and the traffic problems that are commonly encountered today, morning and lunch breaks, and occasional trips to the supply house for installation materials. Getting set up to do the work requires time also—for example, time for bringing the tools and ladders to the area of the building where the work will be done and time for putting them away at the end of the day. I suspect, for the most part, that installers begin to pack their things up before it's actually time to quit, so that they can walk off the job when it's time to go. In other words, the tools are probably packed up on company time even though downtime was not figured into the estimate. The estimate seems to reflect all productive time, and that is why the allowances are as generous as they are, to try to relate *true* production output to the estimate. If the workers in your shop are particularly efficient and devoted, you will be able to factor your labor-hour totals downward, *maybe*. If any downward factoring is to be done in your shop, you should be the one to do it. You might consider the unit labor-hour allowances as conservative national averages, and you might beat them. But controls installation, coordination, and setup take time. The labor-hour allowances listed in Table 3-1 are probably close to being right.

WARRANTY WORK AND TROUBLESHOOTING

Some additional comments on the validity of the estimated labor-hours listed in Table 3-1 are in order. There is certainly a temptation to reduce the total allocated hours for any project at bid time in the hope that by scaling down the time, your bid will become more competitive. If you are an experienced controls contractor, you already know that very often customers lean on you for diagnostic and troubleshooting support when problems arise on an installation during start-up and during the first year of the warranty. Those items pertaining to true warranty work are allowed for on Form 14-1, a sample bid summary sheet, under the heading "Peripheral costs." But in providing diagnostic and troubleshooting support, you frequently find out that the problem is not yours and that you have been called to do the job simply because the customer did not check out the problem beforehand to see who was responsible for handling it. In many cases, mechanical contractors automatically pass the buck to the controls contractor whenever a call relating to difficulty with temperatures is received. If you have been in the industry for some time, you probably have experienced this and have some background for assessing costs related to this type of call. If you trim your labor-hours on a project that you're estimating, you remove a portion of time which, to a very slight degree, is built into the labor-hour units to cover diagnostic work which is broadly related to your total involvement in getting the job going. Such work is not under warranty, since it does not concern your equipment, and it may not have been your work in the first place.

An example will illustrate what often happens to controls contractors when they are interested in satisfying their customers and in placing themselves in their customers' good graces in order to get further work from them. My company had furnished and installed a control system in a new school. The system featured individual heat pumps for each classroom plus heat pumps for the administrative areas. During the cooling season, these air-to-water heat pumps rejected their heat to a continuously flowing water loop. The water returned to the equipment room, where pumps pushed the water out over an evaporative cooler for heat rejection to the atmosphere. A set point of 80°F was established as the control point above which the flowing loop temperature would not be permitted to rise. When cooler weather arrived, the evaporative cooler shut down, and the water passed through without cooling taking place because now the loop was "heat-sourced." The heat pumps delivered heat to their respective areas by extracting the heat from the water, consequently "chilling" the water. Thus, in the heating season, supplemental heat was necessary to maintain a minimum water loop temperature of 70°F. If the water loop temperature fell below 60°F, the system would trigger an alarm, shut down the heat pumps throughout the building, and stop the water circulating pumps.

In October, I received a call from my customer, the mechanical contractor, who announced that the school had no heat and that the controls had everything "locked out." Neither the mechanical contractor nor the school personnel knew what to do. My company had given instruction 3 months earlier to the school's operating engineer, but he was gone now, and no one else was familiar with the system, even though diagrams and a written sequence of operation were posted on the job.

When I got to the school, I discovered that the cap boiler was off. The disconnect serving this 180-KW boiler was in the off position, *and* the boiler manual bypass valve was open, while the boiler manual return and supply valves were closed. The controls had performed exactly as they were supposed to. The water temperature in the loop on this cold day in late October had dropped to 60°F, and there was no means of obtaining supplemental heat. No wonder the school had no heat. All the heat pumps were off, as they were supposed to be in this alarm state. Not knowing why the boiler was off, I called my customer's office, but the people familiar with the job were out. I left word and waited. I didn't want to just turn the boiler on, because (1) I didn't know why it was shut down and (2) it was not my responsibility to energize it, since I hadn't furnished it and it was still under warranty by the mechanical contractor. An hour later, I called back, and this time I talked to the mechanical contractor, who told me that he had turned the boiler off because he didn't see any reason for it to be on during the summer, and that I could turn everything back on. I responded that, no, I couldn't do that. I explained that he needed to come out and take that responsibility, but that I'd stay at the school to get the control system reset when he had the boiler going. Then I waited some more.

When he arrived, he opened the valves to the boiler, closed the bypass, and closed the disconnect for the boiler. But—guess what—the boiler didn't come on. I noticed a paddle-type water flow switch on the inlet spud near the bottom of the electric boiler and decided to "read" the contacts to see if they were indicating flow. They weren't, even though we knew buckets of water were racing through the unit. Adjusting the sensitivity didn't help. The switch would not make contact. So we shut the boiler down, drained it, and removed the flow switch—only to find that at the factory where the switch was installed before the boiler was shipped, someone had neglected to trim the paddle to fit the pipe; when the paddle was screwed into its nipple, the paddle jammed against the bottom of the inlet spud, and the switch could not make contact, no matter how much flow occurred, because the paddle couldn't move. Working together, the mechanical contractor and I trimmed the paddle until it was right and got the boiler going so that I could reset the control system and the school could have heat. But none of the work I did related to my control system, and I lost nearly a whole day of work. I didn't mind lending a hand in these diagnostic and corrective efforts, but the point is that the work I did that day was not warranty work. It was work that I got involved in because the contractor didn't do his job thoroughly. The boiler had never been checked out and had never heated any water before we fixed the problem. It couldn't have. The water flow switch was holding the control circuit in the open position, and the contractors for the six steps of electric resistance heat could not pull in. The water flow switch had been furnished prewired with the electric boiler. And the manufacturer's representative hadn't checked it out. I wound up doing his work for free.

This is not an unusual situation, as any controls contractor will tell you. Controls contractors don't mind giving service like that. They are glad they can help out, but it takes time, and unless the estimate has sufficient time built into it to cover situations such as this, labor costs will run over and eat into job profits.

As the complexity of a job increases, your involvement in troubleshooting will increase, because until you can be sure that your controls are not the source of a problem, your system is suspect. Too often in the controls business, the controls contractors are guilty until proven innocent. In the long run, the customer is appreciative and probably will favor the helpful controls contractor insofar as possible—unless there's enough difference in costs at bid time, and then that favor might be forgotten. The point is that controls contractors need to have a handle on the costs and to know where the costs are coming from. Until you have experience with labor costs and can appreciate the impact that a dependent customer can have on your profit and loss statement, you might be wise to estimate conservatively, allowing adequate time to get the system up and going in concert with the HVAC equipment.

4

Fail-Safe Considerations for Automatic Controls

LOSS OF COMPRESSED AIR AND CONTROL POWER

In estimating control systems and preparing pneumatic takeoff diagrams, it is well to keep in mind the way the system would fail if compressed air were lost for any reason in a pneumatic application. In some instances, when compressed air is lost, it is important for the contact or actuator to be in a specific position. Safeguards must be employed to ensure (1) that a fail-safe reaction will cause the contact or actuator to go to the desired position, and (2) that the automatic restoration of the operation of equipment occurs in a safe and orderly manner when the compressed air is restored.

Electric Heat

Consider pneumatic-electric relays (PEs) that control electric heat. If a controls contractor were to diagram for the use of direct-acting thermostats (a rise in temperature causes a rise in pneumatic output), then the PE would, of necessity, have to be wired normally closed. With a fall in temperature this allows the cold thermostat (low pneumatic output) to remove air from the PE, causing the contact to close and energizing the electric heat. But if all the compressed air were lost for any reason, the PE would close its contact, and the electric heat would remain on. If possible, it is preferable to use a reverse-acting thermostat; with this thermostat a fall in temperature causes a rise in pneumatic output and a rise in temperature causes a fall in pneumatic output. Then the PEs involved can be wired normally open, requiring an increasing pneumatic signal to them to close their contacts. This fail-safe reaction would cause the PEs to open their contacts should compressed air be lost.

Steam-to-Hot-Water Converters

Steam-to-hot-water converters also require careful application of the fail-safe reaction. Generally, normally closed steam valves are used to control the steam going to the converter, and for good reason. When the pneumatic system goes down, this control valve needs to move toward the closed position. If the valve is normally open rather than normally closed, loss of the pneumatic signal causes the valve to assume the fully open position and uncontrolled amounts of steam are introduced into the converter. Eventually, the water side of the converter overheats and produces pockets of steam as the water flashes into the vapor state, causing excessively high temperatures in the hydronic system and water hammering in the piping, along with pump cavitation. These undesirable effects in the hot water piping can damage the piping and the pump. Therefore, it is wise to estimate normally closed steam control valves when they are used in conjunction with control of a steam-to-hot-water converter. In the normally closed system, a reverse-acting hot water temperature controller removes the pneumatic signal from the control valve on a rise in the temperature of the heating hot water supply and an interlocking electro-pneumatic relay (EP) exhausts all air from the diaphragm of the control valve anytime the pump serving the converter is shut down. And a high-limit controller, which may be specified for the converter, merely has to overcall the primary controller and remove the pneumatic signal if necessary to maintain safe temperatures.

Other Fail-Safe Applications

Fail-safe applications at air handlers should always cause the outside air intake dampers to close completely whenever the unit is to be shut down. A fan room unit ducted to the outside air for introduction of fresh air during operation needs to be isolated from outside temperatures if the fan is stopped and the unit temperature controls are prevented from maintaining proper casing temperatures. In extremely cold weather, freezing air migrating into the unit can freeze chilled water coils and heating coils as well. Actuators which spring-return to the closed damper state must be employed. Then, if compressed air is lost, all the fresh air will be locked out and the water coils will be protected from freezing. Fan EPs will exhaust a signal from the motor on fan shutdown to achieve the same result. In order to maintain sufficiently warm casing temperatures, rooftop units may require casing thermostats that will assume control of the heating coils anytime the unit is not running.

DEALING WITH POTENTIAL INSTALLATION PROBLEMS

Estimators of control systems should be alert to potentially troublesome conditions and be prepared to bring them to the attention of architects before installation is accomplished. The more trouble-free that control contractors can make the system for the owners and designers, the easier the job will be in taking care of the system and providing satisfactory control of the equipment being served.

If you are writing the spec, then that is the time to discuss such concerns with the mechanical consultant. But if the job is already out for bids, pointing out potential problems may place you in the position of having to include remedial devices, while your competitors merely quote on the spec as is. However, making a note of a potential problem and bringing it up after you have the contract for installation could well generate an extra job for you. This is not to say that the time to raise questions is after you have the contract and you are assured of an extra work order. But if there is insufficient time for an addendum to be issued to all bidders, or if the designer is reluctant to have the spec tampered with, bringing up potential problems could force you to make an inclusion not required by other bidders.

Another way to look at this situation is that you have an opportunity to help the designers and owners at no cost to them, and by doing so, you are demonstrating your willingness to work with them and provide what they need as a gesture of goodwill. The important point here, though, is that it is never to your credit to make an installation that has an inherent problem of which you are aware. Bring the problem to the attention of those concerned somewhere along the line, and the job will be more successful because of it. I feel that if there is time for an addendum to be issued, most consultants will be glad to issue a corrective directive and to have all bidders look at the same job. Moreover, the consultants will probably appreciate your having helped prevent a troublesome problem.

5

Cautions for Systematic Restart of Control Systems

Restoring control conditions in an orderly manner after shutdown is an important consideration. If an air handler has been off for a while in cold weather, the discharge air temperature controller may be warmer than normal if the heating coil valve assumed the normally open position upon shutdown. Using restrictors in signal lines to outside air damper actuators allows the fresh air intake damper to open slowly. This can be quite important, for without restrictors a warm discharge temperature controller will push open the fresh air dampers as soon as the unit is restarted. Freezestats are often accidentally tripped by superheated discharge sensors. In very cold weather, the sudden opening of the fresh air intake damper upon start-up introduces excessively cold air into the intake side of the air handler and generally causes the freezestat to open its contact to shut down the fan. This occurs for two reasons: (1) the freezestat is more sensitive and reacts more quickly than the discharge controller and (2) the freezestat is usually closer to the fresh air intake than is the discharge controller. Controls contractors should be aware of the need to prevent premature damper opening, and they should make provisions for this in their estimating by noting the need for a restrictor on their sketches. The cost is insignificant. The restrictor must be provided for as a matter of good judgment in the interest of having the best possible control system.

PREVENTING ENERGY PEAKS

Morning start-up can be staggered to prevent energy peaks; however, energy peaks can still occur after a power failure if all HVAC motors are allowed to come back on the line at the same time. When the daytime signal is pneumatic and serves to close the PEs to start the fans, a power failure of very short duration will not deplete the compressed air supply in the compressor reservoir, and so the PEs in the HVAC control circuits will remain in their on position with the contacts closed. When commercial power is restored, the equipment will start up simultaneously. To combat this, the clock-actuated EPs which return systems to daytime status must be energized during the daytime period. Then, a power failure will deenergize all EPs and open all PE circuits. In this way, the restoration of utility-provided power will have the same effect as clock-initiated restoration of normal occupied conditions, and the varying restrictor-controlled branch line feed rates to the PEs will occur systematically.

MAINTAINING TEMPERATURES

Clock-controlled EPs which are deenergized during the day generally allow normally closed PEs to maintain motor control circuits for continuous equipment operation. This is an excellent plan when the line is routed through a night thermostat, for instant let-through occurs at the end of the day when the EP is energized to pass air, as the warm direct-acting night thermostat passes the signal on to the PE to stop the fan. Subsequently, a fall in temperature later in the night to the set point of the night thermostat causes the night thermostat to decrease its pneumatic output and allows the PE to close, thereby restarting the fan in the air handler. Thus, space temperatures are maintained at the setting of the night thermostat by cycling the air-handling unit's fan. To use normally open PEs in the fan circuits, a reverse-acting night thermostat is required, and its PE must be wired in parallel with a clock contact connected to control the fan circuit directly. In this way, when power is restored, the HVAC equipment will return to the on mode in an orderly fashion, with preferred sequential start-up, thus avoiding the in-rush of current that occurs when all equipment is brought back on at the same moment. While it may not be economically feasible to include this option in the estimate, the cost is relatively small, and the owner might decide to have installation done later on in the form of a job extra. When many motors are involved, the savings that are realized by preventing a new billing peak will largely offset the cost of making this improvement. Your vigilance in areas such as this will be rewarded by the respect you enjoy from your customers as they begin to realize they can rely on you to help them operate their buildings more efficiently. Caring for the customer is sure to be recognized as a sign of your desire to provide them with the best system possible.

FAN SHUTDOWN

In systems where static pressure controls are employed, fan shutdown should cause the fan capacity controls to position the inlet vanes closed, or set the blade pitch to minimum in the case of a vane axial pitch controller. Upon start-up, then, the fan motor can reach its normal rpm without being penalized by having the fan fully loaded, which might cause the motor running overcurrent protection devices to react. Operators preferably should be connected so that a loss of compressed air will permit spring-return to the closed or unloaded position. Additionally, EPs should be used to interlock the static pressure controls with fan operation. Consider a reverse-acting duct static pressure controller. On a fall in duct static, the pressure controller output will signal the capacity actuator at the fan to deliver more air. If the fan is stopped, static pressure will drop to zero, and the static pressure controller will call for maximum delivery. Without EP interlock to ensure minimum positioning of the static pressure actuators, restarting of the fan will cause the fan to rev up, with the vanes pitched for highest delivery. Where restrictors have been used to delay the signal to the vane actuator to prevent hunting, the controller signal may not arrive in time to prevent the development of full rpm at the maximum open setting. Casing and duct damage may result if the motor overload or static high-limit devices don't react first.

Sometimes vortex or blade pitch actuators are reverse-connected, where a retracting or spring-returning actuator actually increases fan output, and the actuator shaft must be extended in order to throttle down on air being pumped into the duct system. In such a case, EP interlock may well place a full signal on the control ports of capacity motors to force them to minimum, but protection against system overpressurization upon loss of compressed air must be provided for too. Steps should be taken to stop the fan from running if the air compressors go down while the fan is in operation. This can easily be accomplished through the use of a PE wired into the fan control circuit so that its normally open contacts can interrupt the motor starter holding coil circuit if for any reason compressed air is lost.

STATIC PRESSURE CONTROLLER SWING

Another area of static pressure control work which bears watching is short cycling of the pitch control actuator as a result of static pressure controller swing, or hunting. The use of restrictors to slow down static pressure controller reaction was mentioned in the discussion about returning the vane actuator to minimum upon shutdown. Such restrictors should be installed in the pneumatic signal line connected to the actuator, with an instantaneous check release to provide for instant overriding action by duct static high-limit devices.

Figure 5-1 shows a check valve (or pneumatic diode) connected in parallel with a restrictor. On increasing pressure from the duct static pressure controller, provided that high-limit controller HL-1 is satisfied, the signal seeps through the restrictor (abbreviated "Res." on the figure) and causes the capacity control

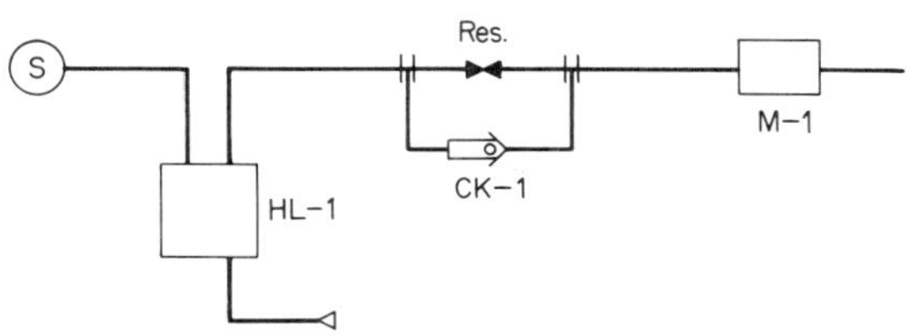

Figure 5-1 Retarded static signal with instant bleed.

vanes at the fan to open slowly. This permits a very gradual pressurization of the duct static pressure and ensures recognition of increasing system pressures by the static pressure sensor and controller, whose output to HL-1 is shown on Fig. 5-1 as an S within a circle. Stabilization occurs before the full signal ever reaches M-1, as the primary controller begins to back off its output in acknowledgment of satisfactory duct static pressure. As a result, equilibrium can occur between the duct static requirements and the position assumed by M-1, thereby eliminating hunting in most cases. At the same time, any overcall action necessary on the part of HL-1 can be immediately introduced, as check CK-1 pops off seat the moment pressures are lower on the controller side. This permits M-1 to spring-return quickly so as to prevent any adverse affects that could occur from a dangerously high static pressure.

Duct static pressure sensors generally are placed well out into the system, at least two-thirds the length of the longest duct or at the end of the duct run. This ensures that there will be sufficient pressurization throughout the duct system to permit proper operation of volume regulators within terminal units and satisfactory control by system-powered thermostats that use the pressurized conditioned air flow to control terminal unit actuators. Both static pressure which is too high and static pressure which is too low should be avoided.

6

Estimating Control System Enthalpy Devices

ENTHALPY DEVICES

Estimating enthalpy control should be approached with a clear understanding of the devices which will form the enthalpy loop. Different manufacturers may provide varying numbers of components for the enthalpy sensing and comparison networks. Combination temperature and humidity sensors are available which transmit a pneumatic signal that relates directly to enthalpy, or the total heat content of the airstream. In other devices, enthalpy is a product of separate temperature and humidity readings from both the return and outside air. Savings in labor costs associated with the combination instrument need to be weighed against what may be a higher cost of the instrument itself. Does the enthalpy sensor cost more than the other manufacturers' separate temperature and humidity sensors would cost? Such price comparisons can now be readily made because various name-brand instruments are conveniently marketed by control wholesalers. Therefore, when you have an option about what technique you can use, such an investigation is in order. The nature of remote readouts relating to enthalpy may dictate your choice. Sometimes a readout in Btu's per pound of dry air is required for local or remote panels. This calls for enthalpy transmitters or sensors whose linear pneumatic output tracks a change in the total heat content of the airstream being sensed within a specified range. A conventional receiver gauge, calibrated to the span that is to be observed, will give this readout.

CHOOSING AN ENTHALPY DEVICE

The significance of ventilating decisions based on enthalpy considerations has become widely recognized in recent years. In addition to meeting minimum fresh air requirements, an air-handling unit should also be able to keep out outside air until (1) it is needed for cooling, (2) it is at a low enough total heat content (enthalpy) to contribute to cooling, and (3) it has a total heat content lower than that of the return air. Too much outside air in the winter adds to the heating load and wastes energy, and too much outside air in the summer adds to the cooling load and wastes energy also. In order for any enthalpy-based ventilation decision to save a building owner money, it must involve accurate devices in good working order with good characteristics of repeatability. Unfortunately, enthalpy and relative humidity are probably the two weakest links in all manufacturers' lines of controls. Choose your manufacturer carefully and seek others' opinions of the brands you're considering. Buy good sensors and controls when enthalpy is involved. Faulty enthalpy devices open fresh air dampers at the wrong time and don't open them when they should. This is an area where building owners can save on energy costs, and you can be sure you will hear from them if fresh air dampers are wide open in hot, humid weather.

Figure 6-1 is a typical enthalpy control diagram that will help show what an enthalpy device is supposed to do. In the figure, EP air is fed to enthalpy controller EC-1 and minimum position switch MP-1 to ensure that all compressed air is removed from the controls whenever the fan stops; this guarantees the closing of the relief and outside air dampers. Temperature transmitter TT-1 and humidity transmitter HT-1 are situated in the return airstream to sense the temperature and relative humidity and provide input information to EC-1. Temperature transmitter TT-2 and relative humidity transmitter HT-2 are placed in the outside air to sense the temperature and

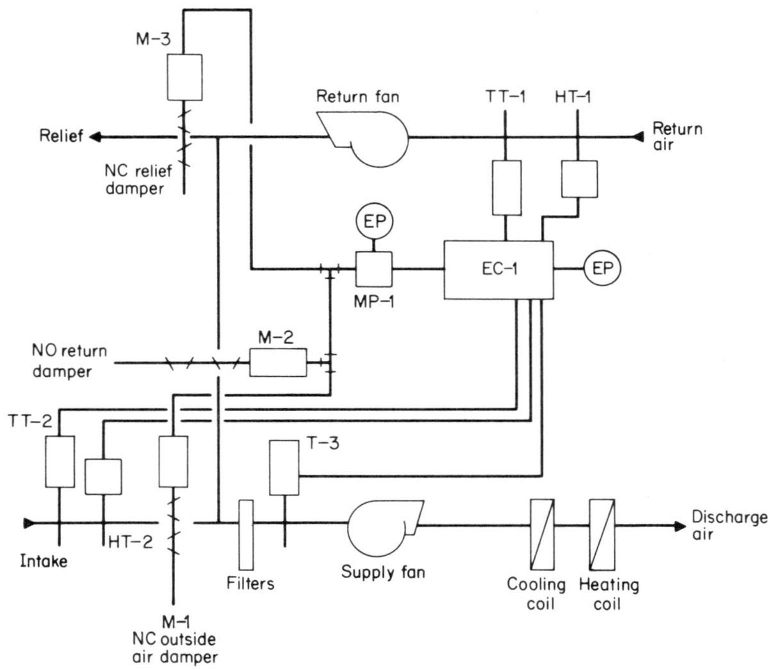

Figure 6-1 Enthalpy control schematic.

relative humidity of the outside air and provide that information to EC-1. Thermostat T-3, located in the mixed air, sends its signal through EC-1 to control the temperature entering the fan. The set point of T-3 might be 55°F, and on the basis of information at EC-1, T-3 modulates open the outside air and relief air dampers and proportionately closes the return air damper so that it maintains its set point. In the absence of a signal from EC-1 to damper actuators M-1, M-2, and M-3, MP-1 positions these actuators for a predetermined amount of fresh air whenever the fan is running in the occupied mode in order to ensure that minimum code ventilation and air change requirements are met. When the calculated total heat of the return air as sensed by TT-1 and HT-1 is greater than the calculated total heat of the outside air as sensed by TT-2 and HT-2, the dampers operate as required to satisfy the 55°F set point of T-3. When the calculated total heat of the outside air is greater than the calculated heat of the return air, the damper actuators are placed under the control of MP-1 only. Thus, in hot, humid weather, when the inside air represents a light load as compared with the outside air, EC-1 saves the owners money. The enthalpy controller looks at the latent heat content of the different sources of air as well as the sensible heat, and makes its decision based on more valid information than would be obtained from just temperature sensors alone.

ASSIGNING LABOR-HOURS

Because of the increased emphasis on sensing enthalpy rather than only air temperature, enthalpy devices will continue to be more frequently used. Equipment room labor-hours can be selected from Table 3-1, which provides labor-hours for fan room devices. The enthalpy controller is basically a comparison device which looks at the total heat content of outside and return air. The source with the smaller heat content is utilized in the summertime to lighten the cooling load. Invariably, a changeover relay is incorporated into the enthalpy control unit, and merely locks out the ventilation cycle once the outside air enthalpy exceeds the return air enthalpy. An enthalpy control unit is not extraordinarily complex; but because of the many signals being introduced into the unit and because of the numerous pneumatic connections required by the unit, sufficient labor-hours must be allowed to satisfactorily commission the unit. Thus, more labor-hours are assigned to an enthalpy controller as well as to receiver controllers, as Table 3-1 shows.

From Table 3-1 the appropriate hours for the enthalpy controller can be selected, as well as hours for the sensors, on the basis of the installation methods to be used. The total number of labor-hours required is then found when all the components attendant to the enthalpy controller plus the enthalpy controller itself are considered. This labor-hour requirement becomes very realistic in the field when the technician is involved in setup and demonstration of the enthalpy control scheme and in verification of the transmitter signals. Setup is quite important.

7

Estimating Controls for Fan-Powered VAV Boxes

The use of fan-powered variable air volume (FP/VAV) boxes in commercial buildings provides flexibility, because at night, during periods when the building is not occupied, the air-handling unit does not need to be restarted just to maintain a reduced nighttime temperature. In a central fan system serving these FP/VAV boxes, each thermostat is able to start the fan within the FP/VAV boxes it controls for local, independent reheating of recirculated air to satisfy nighttime reduced temperature requirements in its area. Thus, instead of having to start what could be a very large air-handling unit with its very large heat source, the fractional-horsepower motors within the boxes are cycled indiscriminately in conjunction with local electric reheat coils.

OPERATION OF FP/VAV BOXES

In examining the operation of these FP/VAV boxes, it can be seen that during normal periods of occupancy, reverse-acting room thermostat T-1 (Fig. 7-1) is indexed to its normal daytime temperature setting by an intentional change in main air pressure at the thermostat. As its space becomes too cool, T-1 increases its pneumatic branch line output to PE-1, PE-2, and pneumatic damper actuator M-1 simultaneously. M-1 reacts first, because its spring range (SR) of 3 to 7 lb is lower than the setting of either of the PEs. As M-1 responds to the increasing pneumatic signal, supply air damper D-1 within the FP/VAV box moves toward the closed position, while return air damper D-2 correspondingly moves toward the open position. Thus, the incoming 60°F air from the air handler is proportionately blocked while the ceiling plenum return air damper is opened to permit the introduction of plenum return air to the integral fan within the FP/VAV box. These two dampers may be staged so that D-2 does not open until D-1 is closed. On a continued fall in space temperature, fan F-1 is started and electric reheat coil RHC is energized for local heating to satisfy the requirements of T-1. When the end of the occupied period is reached, such as at the end of the normal working day, the main air-handling

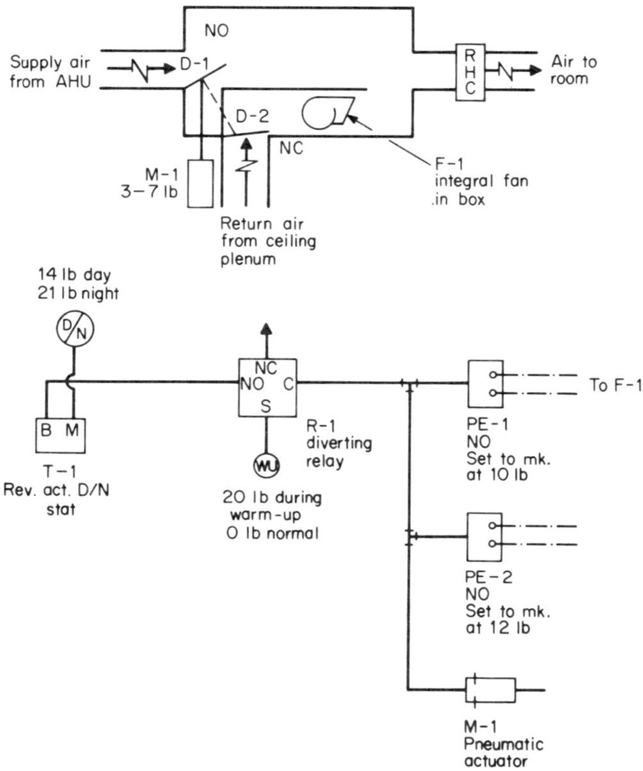

Figure 7-1 Pneumatic control of fan-powered variable air volume box with morning warm-up.

unit is stopped, and all T-1 thermostats are indexed to a lower temperature setting by a change in the main air pressure. On a fall in temperature during periods when the building is not occupied, the T-1 thermostats continue to add reheat when the spaces become cold, just as before, but seek to maintain the reduced temperatures. When the weather is cold, the air handler starts up early in the morning—before the normal building occupancy time—in a warm-up mode, with no fresh air introduced from the outside and with perhaps a main central electric heating coil in the large air-handling unit under the control of a return air thermostat. Thus, during this warm-up period only, the air-handling unit delivers warm air until the return air temperature, which represents the conditions on the floors within the building, reaches perhaps 68°F. Then the unit reverts to its normal daytime operation. At the same time that the warm-up cycle begins, a pneumatic switching signal is fed to diverting relay R-1 at each FP/VAV box, removing the T-1 thermostats completely from the control of the PEs and M-1. This permits all the D-1 dampers to go to the full-open position and closes all the D-2 dampers to ensure that the air being delivered by the main air-handling unit during warm-up is accepted by each FP/VAV box and delivered into the spaces.

These FP/VAV units are used generally in the perimeter areas of buildings. The interior area is usually served by conventional VAV boxes without integral fans, and so during nighttime, the interior VAV boxes are inactive since the main air-handling unit is off.

FP/VAV boxes are finding increasing application because they offer variable air volume control with its attendant savings in energy and because they operate during off periods like a unit heater. Also, the warm-up provision allows the FP/VAV controls to be bypassed until the building has reached acceptable temperatures for normal occupancy.

FP/VAV units are ideally suited for use with optimizing programmers which inhibit the introduction of any outside air until temperatures in the building have reached normal levels and which can vary the length of the warm-up time in relation to outside air temperatures and building heat retention capabilities.

ESTIMATING PNEUMATIC CONTROLS FOR FP/VAV BOXES

In estimating pneumatic controls for FP/VAV boxes, check before bid time with the manufacturer's representative who is supplying the FP/VAV boxes to see if the PEs and pneumatic actuator M-1 are included. When significant quantities are involved, such coordination can help you bid the job accurately and in accordance with what the box manufacturer will provide. Also, verify the action of the thermostat required. Generally, it is reverse-acting, but it can vary with different manufacturers. When control wiring is the responsibility of the controls contractor, check to see if the factory PEs will be prewired and not just mounted. If you are to do the wiring, you may need to allow for this in your control wiring cost.

Some box manufacturers use a control scheme which keeps return air damper D-2 closed completely during daytime operation. In this instance, actuator M-1 might be indexed in a two-position manner along with lockout of PE-1 to the integral box fan. If this control scheme is used, the air volume during the day on a fall in room temperature is reduced to a predetermined minimum air flow, but the air flow is sufficient to permit satisfactory operation of electric reheat coil RHC. A fall in temperature, then, merely adds reheat to the reduced flow of 60°F air from the main air-handling unit; the other boxes may or may not need a maximium flow of 60°F air. Thus, VAV systems are very flexible. A check of the specified sequence of operation on the contract bid documents should reveal the intent of the mechanical design engineer. Then you can fit the appropriate control scheme to the condition required and cross-check the particulars of FP/VAV box control with the manufacturer's rep.

Diverting relay R-1, shown in Fig. 7-1, receives the pneumatic switching signal only during warm-up, and at that time R-1 exhausts all compressed air from the line to the PEs and actuator M-1, while blocking the pneumatic branch line output from thermostat T-1. When one thermostat serves two or more FP/VAV boxes, only one diverting relay need be used with each thermostat; one per FP/VAV box is *not* required. Should you estimate one per FP/VAV box, you will have more than you need and may not win the bid. One per thermostat is all that is required because it is the *common signal* from T-1 that is being blocked, not a separate signal to each box. R-1 can be located above the ceiling at the top of the partition where the thermostat is and connected to catch the branch line as it rises from the thermostat in the wall below. Make sure to estimate the pneumatic warm-up signal tubing to each thermostat location as well as the day-night air main tubing to each thermostat. Then figure the amount of tubing needed to reach each FP/VAV box, totaling the distances from each thermostat to each FP/VAV box.

One word of caution: Some VAV boxes require constant compressed air pressure in order for their built-in VAV controllers to work. Therefore,

1. Verify the requirements for main air to the FP/VAV box.
2. Determine if you can use the same day-night air main tubing that serves the thermostats. The FP/VAV controller might not work at the reduced main air pressure which is used to tell the thermostat to control daytime temperatures. If the controller will not work with the reduced main air pressure, you will need to run another, separate constant-pressure air main to each of the boxes along with your day-night air main *and* your warm-up line.
3. Add the air consumption of the FP/VAV controllers to the air consumption of your other pneumatic devices in arriving at (1) the size of the air compressor required and (2) the size of the air mains.

8

Estimating Controls for Air-to-Water Heat Pumps

The use of air-to-water heat pumps for heating and cooling individual rooms or areas in buildings is increasing. The heat pumps usually use a water-glycol mixture circuit for their heat source in the winter and for their heat sink in the summer. Air-to-water heat pumps are often found in medium-sized buildings, where temperature control flexibility and intermittent area occupancy are factors. The heat pump units themselves come complete with operational and safety controls and are frequently shipped with wall thermostats or sensors for field mounting. However, in a typical water source application, the controls contractor is often involved because of (1) the required mounting and wiring of the wall sensor devices, (2) the required water-glycol loop temperature controls, and (3) the necessity for designing and connecting a safety-dependent control circuit to shut down the heat pumps if the water-glycol loop temperatures exceed preset temperature limits. Since the controls contractor may have to perform these jobs, it will be helpful to examine a schematic of a typical application and become familiar with what to look for and how to estimate the controls required to maintain stable water-glycol loop temperatures and initiate heat pump shutdown when dangerous temperatures exist in the loop. Here "dangerous" means the threat of damage to the heat pump refrigerant circuit from excessive water temperatures.

OPERATION OF AIR-TO-WATER HEAT PUMPS

Because of the way a heat pump operates, it requires a source of heat in order to heat the space it serves and a source of heat rejection in order to cool the space. In the air-to-water installations in today's commercial buildings, the circulating water-glycol loop, maintained at temperatures between 65° and 85°F, is the source of both heat and heat rejection. When space heating is required, each heat pump, whether freestanding or ducted, can extract heat from the same loop that another unit may be rejecting heat into. The circulating water-glycol mixture returns to the pump room, where the mixture passes through an evaporative cooler, which is in either the passive or the active mode. Then the mixture moves toward a hot water boiler, where a three-way control valve does one of three things: it allows the mixture to go through the boiler, or it allows the mixture to bypass the boiler, or it allows some of the mixture to go through the boiler and some to bypass it. The loop is closed, that is, not open to atmospheric pressure, and temperatures are closely maintained within the 65° to 85°F range already mentioned. As a result, not much boiler energy is required, since (1) the work of heating and cooling is being done by the refrigerant circuits within the heat pumps themselves and (2) the required water temperatures in the winter within the loop are quite low compared with routine hot water boiler installations, where hot water temperatures of 180°F are common.

CONTROLS NECESSARY TO MAINTAIN LOOP TEMPERATURES

Figure 8-1 presents both a water flow schematic and a controls diagram for air-to-water heat pumps. The water flow schematic depicts a common loop as it passes through the boiler room and outdoor evaporative cooler. The

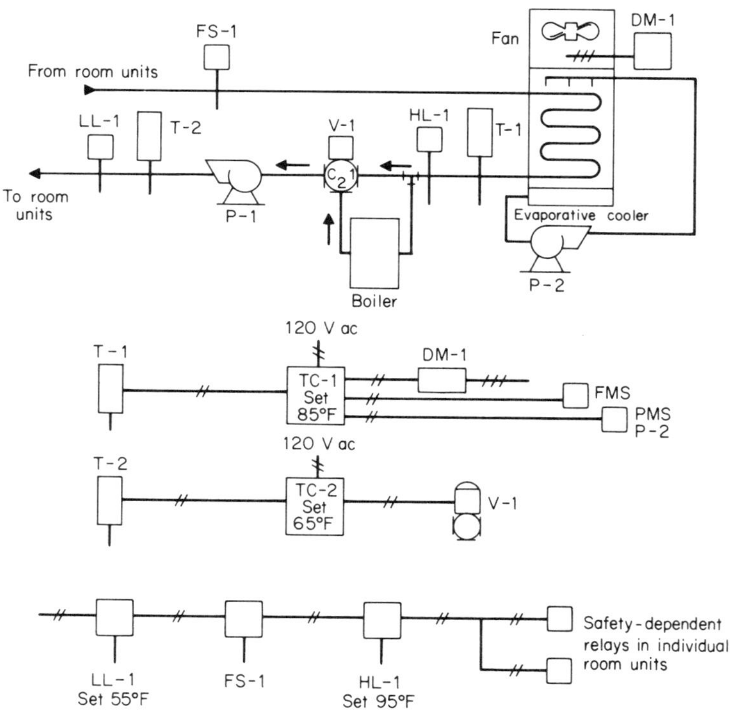

Figure 8-1 Controls for water temperature for air-to-water heat pumps.

controls diagram shows three basic control circuits used in the air-to-water installation. Notice that in the water flow schematic, the pass through the evaporative cooler is through coils of tubing which maintain the closed-loop aspect of the water-glycol circuit. Also shown is the fan, the damper motor and damper, and the circulator at the evaporative cooler. It is important for you to understand the operation of these pieces of equipment thoroughly enough so that you can decide for yourself in an open spec what type of controls to use. In the example in Fig. 8-1, I have chosen electronic sensors and controllers, and I have selected electric devices for the safety-dependent circuit. In order of appearance from top to bottom in the controls diagram, the three basic control circuits are (1) cooling controls T-1 and TC-1, (2) heating controls T-2 and TC-2, and (3) the safety-dependent circuit. T-1 and TC-1 maintain the water-glycol mixture at an upper-level control point of 85°F when the mixture is predominantly heat-laden as a result of a cooling demand within the building. T-2 and TC-2 maintain the mixture at a lower-level control point of 65°F when the mixture is predominantly heat-sourced as a result of a heating demand within the building. The safety-dependent circuit prevents operation of the heat pumps within the building if the mixture is either too hot or too cold or the mixture is not flowing.

In warm weather, the predominant demand within the building is for cool air. On a rise in loop temperature to 85°F, temperature sensor T-1, located in the mixture coming off the evaporative cooler, signals controller TC-1 to open the evaporative cooler damper through damper motor DM-1 to allow natural convection through the evaporative cooler in an attempt to prevent a further rise in the loop temperature. On a continued rise in the loop temperature, TC-1 energizes the evaporative cooler fan motor through energization of fan motor starter FMS. This induced draft assists in lowering the loop temperature, and will probably be successful in lowering it on a day when the outside air temperature is low enough to permit surface-to-surface heat conduction through the tubes within the evaporative cooler. In hot weather, however, on a continued rise in the loop temperature, controller TC-1 energizes the evaporative cooler circulator pump P-2 for circulation of sump water through the spray tree and down across the surface of the tubes within the cooler for evaporative cooling in order to maintain a water-glycol loop temperature of 85°F. On a fall in the loop temperature, the reverse of the above sequence occurs. Naturally, during this process, the boiler is off and the loop mixture bypasses the boiler. Not until all cooling functions of TC-1 cease and the loop temperature falls does T-2 take over.

In cool weather, the predominant demand within the building is for heat. On a fall in the loop termperature, temperature sensor T-2 signals temperature controller TC-2 to modulate three-way control valve V-1 for slight flow through the boiler. At this time, an end switch on the actuator of V-1 most likely energizes the boiler and allows it to operate. On a continued fall in loop temperature at T-2, TC-2 modulates V-1 for up to 100 percent flow (if necessary) of the water-glycol mixture through the boiler to maintain its control point of 65°F. On a rise in the loop temperature, the valve again allows the mixture to bypass the boiler, as required.

Finally, low-limit controller LL-1, water flow switch FS-1, and high-limit controller HL-1 are installed in the pipeline and wired in series so that any of them can open the safety-dependent circuit to stop all heat pumps by deenergizing the controlling relays located within the heat pump units themselves. In such an event, audible alarms are usually energized to attract attention to the fact that the heat pumps are inoperative because of excessive loop temperatures. This will alert maintenance personnel to take corrective action.

ESTIMATING INSTALLATION COSTS

Now that the controls have been examined, it's time to plan for the expense of installing and commissioning the devices. To estimate expenses, four sample estimate sheets are presented in this book. You are already familiar with one—Form 3-1, the equipment room labor and materials estimate sheet. The others are Form 9-1, the space rough-in estimate sheet (Chap. 9); Form 10-1, the control panel estimate sheet (Chap. 10); and Form 14-1, the bid summary sheet (Chap. 14). Once all the forms have been studied, you will be ready to work up an estimate for installing air-to-water heat pump controls. Using this sample job, in Chap. 15 you will develop the entries for the estimate sheets and work them through to arrive at a price for the job. And so keep this heat pump installation in mind since you will be referring to it when you do the estimating exercises.

Going through a mock estimate helps you to unify the approach to estimating controls and lets you become familiar with using the forms. An important objective in the estimating exercises in Chap. 15 is to gather correctly onto the sheets the information that forms the basis for your pricing. Accuracy in estimating is the goal to strive for, and the exercises serve to strengthen the premise that the estimate is an instrument that tells you what to expect in the way of job costs—and profits. When you use the controls installation job discussed in this chapter to do the estimating exercises in Chap. 15, you will be putting into practice for the first time the "how to" of the method in this book.

9

Estimating Space Tubing for Pneumatic Temperature Controls

This chapter discusses the pneumatic system and considers the interconnecting airway in terms of developing an accurate installation cost for use in the preparation of the controls installation estimate.

SELECTING TUBING SIZES

Rough-in can be such a significant portion of a controls job that adequate sizing techniques must, of necessity, play at least a minor role in determining the tubing sizes required for transmission and control. Sizing of air mains is optional on small work but should be routine on big jobs. In large systems, engineering relative to selecting tubing size should be part of the pneumatic estimating process, not because tubing is very expensive, but rather because a large labor savings can be realized through conscientious selection of smaller tubing sizes. Because smaller tubing sizes coupled with cabling can mean much in terms of labor savings in pneumatic work, it is important to be familiar with how adequate size is determined. The successful controls estimator prepares the estimate without tubing size "overkill" and accordingly is placed in a good competitive position and generally winds up with the work. Decisions in the field can lead to the selection of sizes that are not only adequate but also too large, adding to construction costs in both material and labor. Therefore, a properly designed tubing system which originates with the estimator should be carried over, through proper documentation, into the installation phase of the project as well. Every tubing layout that is developed at the time the estimate is prepared should be able to be used for the construction phase of the project, and those involved in the construction phase should feel confident that satisfactory engineering was done initially and that only further verification of tubing sizes is necessary.

You will readily agree that the pressure at the end of a 200-ft length of garden hose will be less than the pressure at the end of a 25-ft length.

Pneumatic systems are no different. For that matter, neither is electric wiring, for extremely long runs introduce voltage drop when the current flow is substantial. Pneumatic piping, be it polyethylene or copper, suffers from the same problem, as long runs take their toll on the flowing pressurized air in the form of skin friction. It is apparent, then, that as a prerequisite for successful system operation, piping with large enough diameters must be selected to prevent inadequate pressures at the pneumatic devices located in the remote sections of the complex under control. But, by the same token, oversizing is wasteful and can rob you of your competitive position when job costs exceed your estimate and force you to compensate on future work. It is better to detail the air piping layout initially and use the advantage of lower costs to obtain the work and prove the validity of the estimate which got you the job.

Air Mains

Consideration of tubing size begins with the air main. Sizing tubes is not hard or time-consuming when the work is properly segmented. Remember that, generally, significant loads, in terms of air consumption, are concentrated in the fan rooms and equipment rooms. By assigning a quantitative value expressed in cubic feet per minute (cfm) of air to each area in which pneumatic devices are concentrated, you can quickly develop a consumption pattern, possibly right on the plans from which you are preparing the estimate. Markings should be made in light pencil which can be erased after the estimate is finished and the information has been transferred to a duplicate for use by the installers at a later date.

CFM Requirements Assuming that all pneumatic equipment has been located on the plans and that the devices have been counted, you can draw a one-line path to all the thermostats and to the equipment rooms. This one-line path represents the main air line feed to the devices and should have lines branching off from it at appropriate points to reach all areas of the building where compressed air is required.

Working from the outer ends, total the air consumption of the pneumatic devices within a given mechanical equipment area in which temperature controls will be installed. This figure, which is expressed in cubic feet per minute, is the required delivery into the segment of the building under consideration. As you work your way back toward the compressor, cfm values cumulate. The air requirement for the area that is farthest away is added to the requirement for the nearest area, and the figure for the nearest area *includes* all cfm requirements beyond that area. Thus, the main air line increases in size as it makes its way back to the source, the air compressor.

Distance and Pressure Drops The cfm requirements are only one of three considerations in selecting proper tubing sizes. The others are distance and pressure drop. In the main air line, each section between points of concentrated load represents an additional length that the air must travel. However, as the air progresses to the outer end of the served structure, less air has to flow. Therefore, each section is sized individually by rating it according to length and cfm and by allowing a predetermined pressure drop within each length. For example, assume for an air supply with 20 lb of pressure that only a 2-lb pressure drop can be allowed at the outer end. First, the total length is measured, and each segment is measured also. Next, each segment is allowed that percentage of drop which its length divided by the overall length represents. Then the drop is plotted against the cfm which must pass through it.

Perhaps an example will help to illustrate the process under discussion here. In Fig. 9-1, the tubing system for a five-story building is laid out in riser form, and the thermostats are segmented and grouped for purposes of ascertaining localized cfm requirements. Each section of main air line is assigned

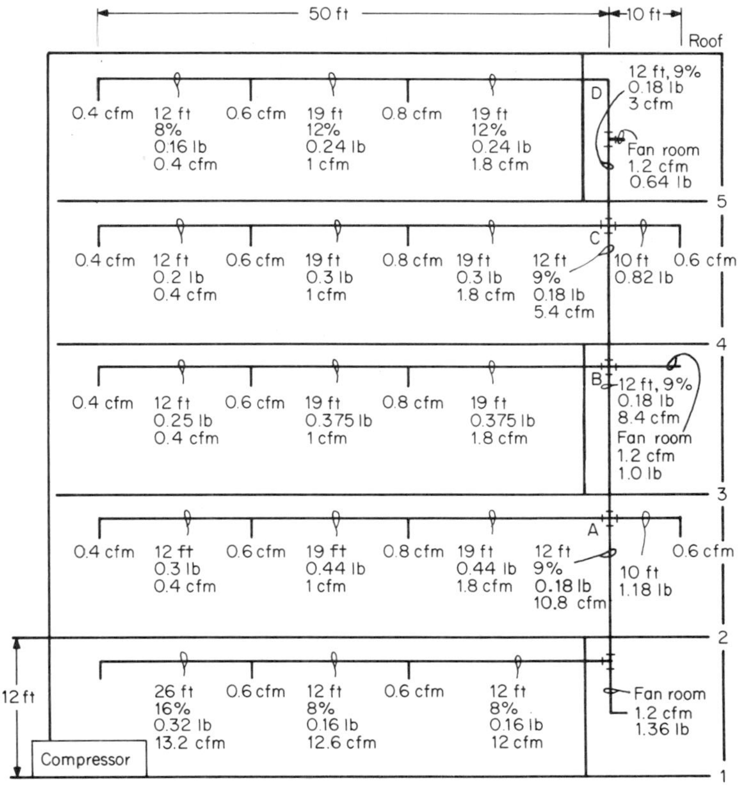

Figure 9-1 Pneumatic system air main riser diagram. Pressure drop at A=0.82 lb, at B=1.0 lb, at C=1.18 lb, and at D=1.36 lb.

a load value expressed in cubic feet per minute as well as a percentage. The percentage is used as a multiplier against the overall pressure drop of 2 lb. Then manufacturers' flowcharts can be consulted for selecting sizes of either copper or polyethylene tubing.

If you look closely at the sample riser in Fig. 9-1, you will find that the longest run is from the air compressor on the first floor to the area farthest from the fan room on the fifth floor. The 12-ft floor-to-floor heights total to a vertical distance of 60 ft. Adding 50 ft of horizontal run on the first and fifth floors increases the distance to 160 ft. From Fig. 9-1, you can see that the vertical riser is to be routed through the fan rooms in order to serve concentrations of devices within those mechanical areas. The longest run in this instance is 160 ft. Taps are strategically located in the horizontal cross-floor runs to serve the space thermostats located on the walls within the areas to be served, and the air flow requirements are noted. The compressed air needs of fan room devices are also totaled and this figure is entered beside the tap for the fan room. Then footages are assigned to each segment of tubing between the junctures of the taps and the main line. Each segment length *of the longest run* is divided by the overall length of 160 ft; the resulting percentage is used to arrive at the allowable pressure drop in terms of pounds per square inch gauge for each individual segment. By multiplying the maximum allowable pressure drop of 2 lb by the calculated percentage of drop permissible within any given segment, an acceptable pressure drop is arrived at in terms of pounds of pressure for the specific length under consideration. Now the cfm requirements of the entire pneumatic system are added up, working from the ends back to the source. This information also appears in Fig. 9-1 on the riser for each section of tubing. Beginning on the fifth floor, the last piece of tubing must pass only 0.4 cfm. The next piece of tubing closer in must pass 1 cfm, the total of the amounts to be delivered through the two taps it feeds. And still closer to the main vertical run, the final space thermostat segment on the fifth floor must pass 1.8 cfm. The vertical segment between floors four and five is required to handle 3 cfm, the 1.8 cfm of the fifth-floor space thermostats and the 1.2 cfm of the fifth-floor fan room.

Each floor is approached in the same way, working back toward the compressor, and the closer the tubing is to the compressor, the more cfm the tubing riser must handle. At the compressor, 13.2 cfm is required. No fan rooms exist on the second and fourth floors; instead, there is usable space in which space thermostats requiring 0.6 cfm are located. Because of a concern for maintaining sufficient pressure at the remotest air-consuming device, an area of greater tolerance is created for the floors below, where horizontal runs take off at points *A*, *B*, and *C* to serve pneumatic controls. Within the longest run, a pressure drop of 1.18 lb exists at point *C* when 5.4 cfm is flowing through the segment between the third and fourth floors. Thus, for the horizontal cross-floor run to the left of the vertical riser on the fourth floor, there can be a total pressure drop of 0.82 lb. All the air required for the fourth and fifth floors has been delivered to point *C*, and there has been a drop of only 1.18 lb. Therefore, the farthest device on the fourth floor can be served with tubing in which an additional 0.82 lb drop can occur. And to the right of the fourth-floor riser, the 10-ft section delivering 0.6 cfm of air can also sustain a drop of 0.82 lb. Thus, segment pressure drops on the second through fourth floors differ according to the remaining differential between the existing pressure drop at the floor juncture and the allowable 2 lb. At the third-floor juncture at point *B*, there is only 1 lb of drop, and so an additional 1-lb drop can occur on either side of the riser. At the second-floor juncture at point *A*, there is only 0.82 lb of drop, and so an additional 1.18 lb of drop can occur on either side of the riser. The amount of pressure drop in the horizontal cross-floor runs on the second through fourth floors can be determined by dividing the cross-floor length of 50 ft into the individual length of the segments, which produces a percentage applicable to the allowable cross-floor pressure drop. The percentage is then calculated for the total permissible pressure drop remaining for that floor, from where the run connects with the riser at points *A*, *B*, and *C*. Finally, each segment is sized according to its length, its cfm requirement, and its allowable pressure drop.

Certain fittings must also be figured into the calculations of pressure drop. Copper solder and compression fittings do not reduce the inside diameter of the tubing and consequently do not have to be figured in. Barbed fittings, however, while being exceptionally convenient to use, do introduce some resistance into the line because they are inserted into the end of the polyethylene tube and thereby, whenever they are used, reduce the tube's inside diameter to the diameter of the fitting. Generally, 10 ft of equivalent tubing length is assigned to each barbed fitting and is added to the segment of tubing that is upstream from the fitting. Cfm requirements differ from one controls manufacturer to another, and because of continuing improvements, newer devices require less air. It is important for controls estimators to be familiar with the pneumatic requirements of the instruments they propose to use and to size the distribution system accordingly. When planning the main, it is important to be cognizant, too, of any compressed air requirements that other equipment served by the controls air compressor might need. Usually, the controls specification details what other equipment must be served with air, so that the controls estimator can plan accordingly.

Average pneumatic control systems might require a normal continuous flow on the order of 5 or 6 cfm per 100 air-consuming devices.

SAFETY CONSIDERATIONS

Fire safety is a concern when instrument-grade polyethylene tubing is used instead of copper tubing in suspended ceilings that serve as return air plenums. Polyethylene pneumatic tubing is commonly used and provides a cost savings to both the owner and the controls contractor. However, in ceiling return air plenums, where emissions of smoke are likely to occur in the event of fire, plastic tubing that is burning may release noxious gases. To ensure that products used in ceiling returns do not contribute to the generation of noxious smoke, building codes have evolved which seek to limit the formation of noxious gases that are produced by the burning of most materials. In this regard, fire-retardant plastic tubing is routinely used in ceiling air plenum spaces and is generally acceptable to most fire marshals as long as the installation does not exceed a predetermined density. Therefore, the guidelines specified in the building code must be followed during the estimating period to ensure that the right tubing will be used when the job is being installed. Even though the controls contractor knows the code requirements, it is a good idea to consult with the fire marshal's office ahead of time to solicit approval of the methods to be employed. Such preliminary contact paves the way for acceptance later on when the representative from the fire marshall's office is inspecting the installations.

Generally, tubing constructed from fire-retardant material bears the letters "FR" to clearly denote its flame-retardant qualities. Quite often, this is of paramount importance to the fire marshal, since it is the one way that the fire marshal can be sure that the tubing that is installed has been approved for the purpose. Appreciate the fire marshal's position, if you will. The purpose of the building code with which you must comply is to protect the lives of others, and the fire marshal is vested with the responsibility of enforcing the code. Human life rests upon the fire marshal's shoulders, and an oversight or omission on the part of the fire marshal could jeopardize others. So you must not feel maligned when an inspector scrutinizes your job. If you do your job properly, there won't be any trouble anyway. It is the estimator's responsibility to use products that meet the requirements of the job. FR tubing must be selected for ceiling return air plenums and applied within the density limits of the local codes.

Care must be taken, as well, in installing thermostat leads. Eight-foot leads of bundled two-tube plastic are commonplace. These prefabricated assemblies make it easier to fish the two tubes into block walls or stud partitions, with excess assembly length exposed above the ceiling line. However, as much as 3 ft might be left protruding from the wall into the plenum cavity. If the thermostat leads are not clearly marked FR, don't be surprised if the inspector calls you down for installing unapproved tubing in the ceiling space. Determine ahead of time whether or not the polyethylene leads that you plan to use bear the proper FR marking. If they do not, their use may be restricted to in-wall applications only. However, in this case, if you estimate properly, you can plan on cutting the leads so that no more than 6 in extends from the wall into the ceiling; this generally will satisfy most inspection officials. These unmarked 6-in leads are then connected to tubing marked FR in order to minimize unmarked tubing. Since only tubing marked FR is acceptable, scale your above-ceiling plenum runs hard to the wall, and you will have estimated enough properly marked tubing to fulfill the job requirements. Looking ahead is so much more economical than looking back. Sufficient planning in this area makes for a smoother job and ensures that you won't run into trouble with code requirements governing nonmetallic pneumatic installations. Contact the local fire marshal's office when in doubt.

SELECTING TUBING SIZES FOR LARGE INSTALLATIONS

From an estimator's standpoint, sizing the tubing for mains in pneumatic systems is usually not a problem unless significant runs of main throughout large areas are needed. For large installations, some engineering techniques may be needed to determine how large the main should be in certain segments of the building. All good estimators should be familiar with the techniques for determining the size of mains when many devices downstream are to be fed through a common tube. This is mandatory in the case of a project such as a multistory office building or an extremely spread-out campus facility where compressed air is to be fed from a central source. For polyethylene tubing, maximum low-pressure main sizing should be 1/2 in. If a larger size is needed, distribution of high-pressure main becomes more desirable, with local pressure-reducing stations strategically located throughout the complex. A maximum of 1/2-in polyethylene tubing is established because of the ready availability of tubing and fittings in this size and because of the desire to minimize stocking sizes with an outside diameter (OD) larger than 1/2 in. As a rule of thumb, compression-type fittings with tubing internal diameter (ID) reinforcing ferrules and nylon compression rings are generally selected for pneumatic pressures above 25 lb, and barbed fittings are satisfactory for pressures 25 lb and below. This means that, for the most part, barbed fittings will be used almost exclusively throughout the major portion of commercial temperature control projects since 25 lb is the highest pressure required for most temperature control devices. Indeed, normally the fittings allowance is built into the tubing cost per foot; the estimator does not need to decide on fittings during the estimating portion of the project, but noting industry practices at this time is nonetheless appropriate.

Figure 9-2 is an example of a campus-style installation and will help to reemphasize some important considerations. The main from the compressor to the farthest point served is considered in segments, and the major intersections seen in the line represent benchmarks of increased capacity requirements. Assume that five thermostats in a remote wing represent the outlying area to be served by the compressor. If each thermostat consumes 0.02 cfm of compressed air at, say, 20 lb of pressure, then 0.1 ft^3 of compressed air is required for the wing. Therefore, the farthermost segment of main must be able to pass 0.1 ft^3 of air as required over the length of the segment. The sizing then is based on the segment that traverses the area within the remote wing, and it is only necessary to determine that the tubing size for that run is sufficient.

Remember, as you move closer to the compressor, you encounter additional pneumatic equipment with additional cfm air flow requirements that will be *additive*. The next segment of tube run is sized to pass the amount of air flow

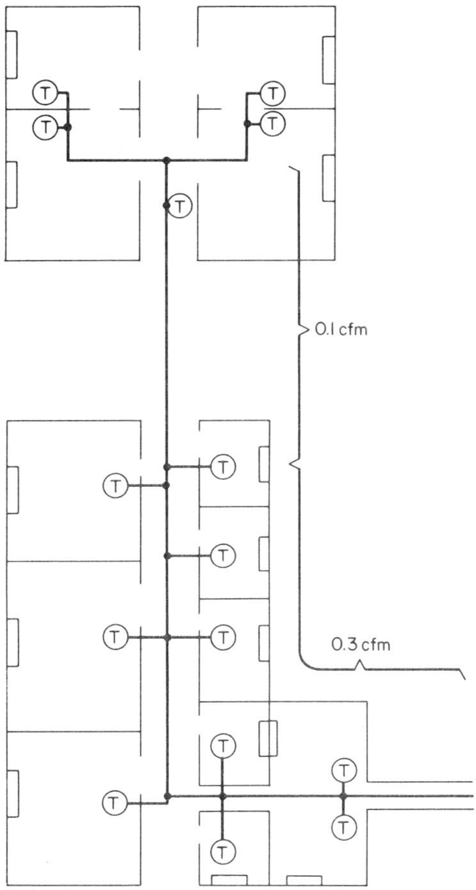

Figure 9-2 Layout for sizing air mains.

it will need plus the 0.1 cfm needed for the previous segment. For example, if the pneumatic load in terms of the cfm requirement for this next segment is 0.2 cfm, then the segment must pass 0.3 cfm—arriving at this amount is just a simple matter of adding the cumulative total of the previous areas to the next area in line. Thus consumption can be added up back toward the compressor to arrive at the largest portion of tubing carrying main to the air-consuming devices, which will, of course, be the first section off the compressor storage vessel.

Figure 9-2 shows a plan view of tubing laid out, with the pneumatic consumption noted next to each segment of the tubing to be installed. The layout is certainly similar to a riser, except that it is presented here in plan view. Each of the remotest segments combines to arrive at the total cfm requirement which the primary line off the compressor storage tank must carry. Then distances are measured to verify how much of the larger-size tubing will be required on the job. If the amount is of sufficient significance, inquiries about pricing for given quantities of the larger size should be made. However, small sections don't demand special consideration.

In the simplified example in Fig. 9-2 the five room thermostats constitute the 0.1 cfm load (each one uses 0.02 cfm). Thus the remotest segment is no more than a feed for five thermostats, with nothing beyond. The adjacent portion of the building, however, has 10 thermostats, each at 0.02 cfm, for a total requirement of 0.2 cfm for those 10 thermostats by themselves. When that total requirement is added to the 0.1 cfm needed for the five thermostats farther on, the requirement for 0.3 cfm is apparent. This approach is repeated throughout the entire project in order to obtain the total requirement at the compressor. The segmenting technique forces the estimator to establish zones within which tubing size must be decided. Note in Fig. 9-2 that the 10 thermostats are tapped at various locations along the route of the main through their portion of the building. But a clear area boundary is evident at the portico connecting the outer wing, and this is where it must be decided whether to recognize a new zone in terms of main air service. The selection of boundaries for the determination of zone main air sizing is not always apparent, and so the estimator is frequently required to examine the bid documents with an eye toward concentrating equipment and grouping space thermostats in a way that will permit air-consuming devices to be grouped for the purpose of establishing cfm requirements along the route of the air main serving the controls. The important point to recognize is that the air main is required to deliver more air to places close to the air compressor and must be sized accordingly. Zoning by load is a practical approach to increasing the diameter of the main air tubing in accordance with how much air the tubing must pass. When this approach is systematically applied, the tubing will prove to be sufficiently large for the controls it has to serve.

The next step is to consult tubing manufacturers' tables to determine what sizes are to be used. Most manufacturers offer a chart that tells how large a tube must be to pass a given number of cfm over a designated distance. When the sizes have been selected from this chart, they can be noted on the layout plan, if one is to be developed.

Allowable Pressure Drops

The basis for choosing tubing sizes is the need to deliver a sufficient quantity of air at strategic locations within the pneumatic system while sustaining acceptable pressure drops. In accordance with established rules for determining the size of tubing in either copper or polyethylene, you must consider the overall length of tubing in any given segment and analyze it for air flow in terms of the allowable pressure drop. The three values that come into play are cfm, distance, and pressure drop. In considering allowable pressure drops, remember that for larger initial pressures a greater in-line pressure drop can be accepted. For a straight main system which requires 20 lb at the instrument, you could conceivably start out with 22 lb and accept a 2-lb drop over the entire length of the system. In such a manner, the various predetermined segments of tubing would be allotted their respective allowable portions of the overall pressure drop.

Assume that you are installing 1000 ft of copper tubing. In looking at the segments that you wish to deal with, you discover three good points along the way where the tubing can be interrupted and judged to be a stand-alone segment. For purposes of this assumption, this creates four segments of 250 ft in length. In an installation where a 2-lb drop is acceptable overall, this means that you can accept a pressure drop of 1/2 lb in each segment. This 1/2-lb drop will occur as the required amount of cfm for devices farther down the line passes through this segment. Thus, the size of the tubing is critical to the preservation of acceptable air pressures at the remotest points in the system. With an eye toward guaranteeing a minimum pressure at the longest run, you can ensure that, at any given point along the way, sufficient air will be available for other devices when you have added in their cfm requirements and held your calculations to acceptable pressure drops within any one segment.

In approaching the task of sizing the pneumatic tubing, attention must be focused on the main air system and the longest single run in that system. For purposes of the calculations here, the longest run will be the distance from the air compressor to the remotest air-consuming device. Once that overall distance is known and the pneumatic air-consuming devices are grouped in acceptable assemblies, then, within any given segment of the tubing, a cfm requirement can be assigned.

Undersizing

The most popular and perhaps the most widely used method for sizing tubing considers each segment in terms of air flow, length, and pressure drop when a cfm requirement has been ascertained. The biggest problem in sizing tubing for pneumatic systems is that the lines may be undersized. When this happens, devices at the remotest points don't get sufficient quantities of compressed air, and they can't operate correctly. Thus, undersizing can be the biggest deterrent to successful operation of a system, while oversizing can result in problems for the estimator when large quantities are involved, since this tends to increase material and installation costs.

For most jobs, the estimator decides how the longest run will be divided. During the takeoff, the estimator can observe clusters of devices and density of equipment in certain areas that tend to create readily determinable segments of pneumatic piping. And during estimating, there is a real opportunity to determine the starting and ending points of different segments. Then, if the estimator is successful in obtaining the job, the information obtained during the estimating process can be carried over into the installation phase of the job, saving time for the construction coordinator who is laying out the pneumatic system.

More often than not, during estimating, sufficient attention is not paid to the sizing of the pneumatic lines. Thus, usually only after the job is ready for installation is the task first encountered by the coordinator or project manager who is developing the installation plan for that project. Good estimators need to be cognizant, however, of the limitation of certain sizes of tubing when large quantities of devices are required over significant lengths. For without sufficient consideration at bid time, the size of the tubing might be underestimated and therefore also the cost of materials and labor. This is particularly true for the larger sizes of copper tubing.

As mentioned before, when it is necessary to use tubing larger than 1/2 in, it may be best to use high-pressure mains routed to strategic locations where pressure-reducing valves can then be set. Then, the low-pressure tubing will originate at the pressure-reducing valve, since the high-pressure line is sized independently of the lower-pressure subfeeders.

Markings on Bundled Tubing

I cannot overemphasize the importance of using polyethylene tubing that is marked FR. When local codes require fire-retardant tubing, it is imperative that the tubing and its jacket, when bundled, be marked in accordance with recognized approved markings. Generally, the individual tubes are adequately marked. Bundled tubing, however, is more likely to present a problem.

In bundled tubing, a jacket is placed over the individual tubes to provide for additional protection and to hold the tubes together as a unit. The jacket itself must offer the same fire-retardant properties that the tubing does. Usually the jacket, as well as the tubes, will be marked to show it is fire-retardant. Occasionally, however, inspectors pick up sample tube cuttings on the job for comparison, only to find a discrepancy between the markings on the jacket and the markings on the tubing within the jacket. This can happen when a large controls manufacturer purchases tubing on a large scale from a tubing manufacturer and prints special markings on the outside jacket that cover up the tubing manufacturer's nationally recognized markings identifying it as a FR-type tubing. A controls company's part numbers or distribution center markings are of no help to a fire marshal, who is looking for readily identifiable marks to indicate that the tubing is fire-retardant. The tubing manufacturer's fire-resistant classification is the one marking the fire marshal particularly wants to see. If unrecognizable marks for special in-house purposes have been used, the matter should be taken up with the fire marshal before the installation is begun. Supported with certification from the tubing supplier, the controls contractor can demonstrate the tubing's suitability and obtain approval ahead of time, both of which contribute to a smoother-running job.

ESTIMATING TUBING

At this point it is necessary to look at the considerations that are involved in planning work in space. The tubing specification generally details the material to be used in the piping system and the way it will be installed. Considerable latitude is allowed today in the choice between metallic and nonmetallic piping, and with good reason. Instrument-grade polyethylene tubing has enjoyed significant success in its applications to pneumatic control systems, and because it costs less than copper tubing, it has saved owners money. Polyethylene tubing can be applied broadly throughout a building, except where temperatures will rise above 150°F, since extreme heat may cause it to soften, making it more susceptible to bursting. However, its use is precluded in limited situations, for example, where it would come into contact with steam piping, boilers, or hot water converters. Therefore, fire-retardant polyethylene tubing is usually named in the specification as an acceptable alternative to copper.

In all likelihood, exposed installations above lift-out ceilings will be permitted, with protective raceways being required where possible mechanical damage could occur, such as down masonry walls or around equipment. Diligent attention is called for in the interpretation of the tubing specification, because much of pneumatic temperature control work is associated with space rough-in. This work consists of installing the lines which must traverse the occupied spaces of the building above ceilings, connecting thermostats, and linking transmitters and switching signals with readout devices and changeover relays. Space rough-in, along with the setting of thermostat heads, is the first part of the installation of a job, and it is the work which will ultimately be concealed by the finished interiors.

Determining Distances

To estimate tubing, distances must be determined, and scaling wheels are a valuable tool for measuring the runs through a building. Begin by determining the path that the main air will have to take to feed air-consuming devices. Take into account all thermostats installed in space, panel-mounted equipment in fan rooms or boiler rooms, and unit-mounted controllers and actuator positioners. Once you have determined the route, do the sizing next in order to discover where the main air tubing needs to be of significantly larger size than the standard 1/4 in. Here the scaling wheel is particularly convenient since measurements of the main air route can be accumulated and read when finished as a total footage for the main air tubing within the building. Working from the control diagram sketches developed during the takeoff, check off each schematic line on your sketch after scaling so that you can be sure you have measured each run. Do this without marking the mechanical contract documents; then the next person who uses them won't be distracted by your marks.

Running Lines in Parallel

When tubes serving different functions run parallel to one another along the same path, it is almost always better to use bundled multiple tubes rather than single lines. Besides the fact that bundled tubes give a neater appearance than a group of single tubes, the protective outer jacket manufactured around the polyethylene tubes meets the requirements of practically all HVAC designers, who insist upon some measure of additional protection for the bare tubes, and the economy of bundled tubes installation is an attractive consideration for the controls system estimator.

For example, assume a job requires a dozen 1/4-in polyethylene tubes traversing a distance of 100 ft. The labor cost savings in a bundled installation are obvious. The additional cost of the bundled assembly over 12 single lines will be more than offset by the cost in labor of pulling 12 single lines. The wise estimator plans on using the same bundle beyond the point where lines begin to drop off to serve various controls. The outer jacket is easily opened, and the proper numbered tube is brought out to connect with a single line extension for an instrument in the area. The unused segment length of the remaining severed tube continues on as a spare. This turns out to be a more economical alternative than stopping the run of bundled tube when the first breakout to an area in a different direction occurs. Up to four or five tubes in the bundle can be allowed to go unused before installation costs become equivalent to materials costs, necessitating the return to single-line tubes. You can make the installation of one bundled run serve the purposes of many tube runs in this manner, reduce costs, and contribute toward the appearance of a neater job too.

With an eye toward the conduit systems required in an electric control installation, keep in mind as you begin planning your conduit runs that:

1. A common conduit with multiple wires is more economical than multiple conduits.
2. Derating of conductors is not a factor with control wires when approved limited current circuits are involved.

You can estimate electrical conduit through the spaces of a building by placing over the floor plan a transparent paper overlay that shows the locations of the thermostats and sensors. On the overlay, draw a line from the equipment room to the first thermostat or sensor to which you must run. Use hash marks to signify numbers of conductors within the raceway. Draw a line to the next thermostat or sensor, which may require two more wires. Now you have four wires to the first sensor, but only two to the next. Add hash marks to the common conduit as you continue out into the building to pick up other devices. Eventually, you'll have a one-line layout of conduit with varying amounts of wires in different sections of the raceway system. Now you can size sections of the common raceway on the basis of the numbers of conductors specified in the tables that appear in codes which dictate fill allowances for different sizes of conduit.

Caution is advised when mixing conductors for various types of sensors or thermostats in space. Occasionally, shielded wire or a separate raceway system is needed for some types of electronic devices. For low-voltage electronic sensors which are very susceptible to interference from other alternating current sources, take care to estimate an individual raceway where necessary and/or shielded wire if needed. Develop the necessary raceway systems, and size your conduits according to the numbers of wires within. The electric system is as easy to enter onto the space rough-in estimate sheet as the pneumatic version is, for the required materials are written out as you check off the sections from your hand-drawn overlay, which is to be kept in the job file.

THE SPACE ROUGH-IN ESTIMATE SHEET

Form 9-1 is a space rough-in estimate sheet. On the form there is a place to enter the name of the job, the date of the estimate, the name of the estimator, and the page number.

Per-foot Pricing of Labor and Materials

The area or floor of the building involved is listed in the column headed "Area or floor." The next two columns provide space to write in the number of feet needed and the size of the materials involved. This applies whether the materials are EMT and wire, rigid conduit and wire, EMT and tubes, bundled tubes, or bare tubes. The format of Form 9-1 permits you to list a completed conduit and wire section as a unit. For example, if your entry is a size of raceway or tubing bundle with wire or tubes included, you might list 60 ft of 3/4-in EMT with 20 no. 14 copper conductors with THHN insulation. The costs for raceway, wire and tubing are determined by the foot. In the example just given, the pricing allowance per foot should cover 3/4-in EMT with 20 conductors inside plus a predetermined add-on per foot for hangers and fittings.

Many shops already have developed unit pricing for sectionalized estimating of completed raceway systems. This method is faster. The hours of labor per foot of installed section are entered with the same idea of unit pricing in mind. The hours of labor per foot are multiplied by the total number of feet to give the total number of labor-hours (the last column in Form 9-1). The same method applies to pneumatic systems. EMT with tubes in place, or bundled tubing with a certain number of polyethylene tubes within, is measured, and the length and size are multiplied to determine cost and hours, as done above for electric system sections. You must determine your own per-foot price of a piece of raceway with wires inside and the labor-hours per foot to install it, or the price and labor-hours per foot for pneumatic raceways with tubes if these are what you're estimating. Base your decision on what you know about production per foot of completed raceway or tubing system from past jobs, or refer to Table 9-1, a chart listing space rough-in hours that appears later in this chapter. Table 9-1 lists labor-hours for various combinations of tubes or wires installed using varying methods. It is discussed in more detail later on.

Space Rough-In Estimate Sheet

Job name ______________________ Date ____________

Estimator ______________________ Page ______ of ______

Area or floor	Feet	Size	Material	Cost/ft, $	Total cost, $	Labor/ 100 ft, h	Total labor, h
Total							

Transfer space rough-in material costs to bid summary sheet, line 6
Transfer space rough-in hours (X 1.3 for high work) to bid summary sheet, line 9

Space devices, stats, sensors, etc.

Area or floor	Qty.	Model	Cost/ea., $	Total cost, $	Labor/ea., h	Total labor, h
Total						

Transfer space device controls costs to bid summary sheet, line 1

Transfer space mounting hours X 1.2 difficulty factor,* if appropriate, to bid summary sheet, line 10

*Ladder required

Space device, new work: connect, mount, and calibrate, 1 h each
Space device, old work: connect, mount, and calibrate, 1.2 h each

Form 9-1 Space rough-in estimate sheet.

For determining the price of material per foot, calculate the cost of 100 ft of EMT and then add a factor for fittings, hangers, and boxes. You may have a feel for what fittings, hangers, and boxes will cost. If not, consider 20 percent to cover them. Then, divide by 100 and add in the per-foot price of wire times the number of conductors involved. Look at a pneumatic installation of a raceway system the same way, and add in the price per foot of tube times the number of tubes involved. As you go on, think in terms of keeping a list of per-foot

Table 9-1 • Space Rough-in Hours Chart*

Qty., Lines	Pipe Size, in	No. 14 THHN Stranded EMT Exp.	EMT Con.	Rigid Exp.	Rigid Con.	EMT Size, in	Polyethylene in EMT Exp.	Con.	¼-in-OD Tubes Polyethylene Bundle Exp.	Con.	Bare Copper Exp.	Con.
2	½	8	6	14	12	¾	7	5	4	2	8	4
4	½	8.2	6.2	14.2	12.2	¾	7.5	6	4.25	2	16	4.5
6	½	8.4	6.4	14.4	12.4	1	13	9	4.5	2.5	20	4.75
8	½	8.6	6.6	14.6	12.6	1 ¼	16	12	4.75	2.5	24	5
10	½	8.8	6.8	14.8	12.8	1 ¼	16.5	14	5	3	28	5.5
12	½	9	7	15	13	1 ¼	17	15	5.25	3	32	6
14	¾	12	10	20	18	1 ½	18.5	16.5	—	—	36	7
16	¾	12.2	10.2	20.2	18.2	1 ½	19	17	—	—	40	8
18	¾	12.4	10.4	20.4	18.4	2	20	18	—	—	42	9
20	¾	12.6	10.6	20.6	18.6	2	21.5	18.5	—	—	44	10
22	¾	12.8	10.8	20.8	18.8	2	23	19	—	—	48	12
24	¾	13	11	21	20	2	25	19.5	—	—	52	14
26	1	16	14	25	24	2	27	21	—	—	54	15
28	1	16.2	14.2	25.2	24.2	2 ½	28	24	—	—	56	16

* Values are in labor-hours per 100 ft.

costs for different sizes of EMT and rigid conduit with various quantities of wire involved in varying multiples of 2. Use the same approach for pneumatic tubing. For five copper lines all running in the same direction, your material costs per foot will be the price per foot of copper times 5. Before long, you will be quite comfortable with this approach to unit per-foot pricing of labor and materials for varying sizes of electric or pneumatic transmission systems.

The entries for the upper portion of Form 9-1 come straight from the scaling you did to determine the distances, and the hash marks will tell you how may wires or tubes are needed in each segmented run. The unit per-foot pricing approach lets you almost picture the transmission systems in the building right on the space rough-in estimate sheet. And it fits nicely into segmenting installation plans that are designed to appraise installer efficiency and develop useful data. Viewing the segmented portions of the space work in this way helps focus your attention on larger runs of conduit and wire or tubing in various portions of the building, thereby highlighting potential trouble spots where congestion may be a factor as the work of other trades converges.

Pricing Control Devices

In the lower portion of Form 9-1, the control devices that are to be used in space are listed by area or floor along with quantity and model number. Next to each listing, the price for a single control is entered and then multiplied by the quantity to obtain the total cost. An hour each for connecting, mounting, and checking each control device, whether it be pneumatic or electric, is recommended; and if a ladder is required to reach the mounting point, as might be the case with above-ceiling plenum thermostats, relays, and damper motors, each hour is to be factored by 1.2. Remember, there may be control dampers or smoke dampers out in the runs of duct above the ceilings in the building. Static pressure sensors in remote duct runs are routinely encountered also. Installing these will require a ladder, and factoring by 1.2 takes this into account. Finally, the space rough-in materials and hours and the space device materials and hours are transferred to the appropriate lines on the bid summary sheet (Form 14-1, Chap. 14). Remember, fittings, hangers, and boxes need to be included in the per-foot costs of the space rough-in materials.

Tubing Fittings Allowances

It is impractical if not impossible to account for every fitting in the pneumatic system at the time that the estimate is made. For this reason, per-foot tubing pricing incorporates an allowance for tubing fittings. For low-pressure polyethylene tubing (25 lb and below), the price per foot includes an allowance for barbed fittings, couplings, tees, right-angle ells, connectors, and caps. For high-pressure polyethylene tubing (above 25 lb), the price per foot includes an allowance for compression fittings, with nylon ferrules and ID inserts for converting the standard metal ferrule compression fitting for use with polyethylene tubing. The price per foot of copper tubing includes an allowance for solder and compression fittings. Therefore, the purchase of fittings to be stocked for use as required can be charged to an installation expense account, thereby avoiding the necessity for accounting for every individual fitting, an impractical task anyway. Under ordinary circumstances in straight-stick commercial temperature control work, this allowance technique works very well, and currently is thought to consist of 22 percent of the per-foot cost for polyethylene tubing and 18 percent of the per-foot cost for copper tubing. Such allowances must, of necessity, be subject to change, owing to the fluctuating prices of copper and brass, but could be reasonably accurate for some time, as inflation increases the cost of the raw materials for polyethylene and copper too. Periodic updating of tubing costs is essential for accurate estimating.

Labor-Hour Allowances

In estimating space rough-in, it is usually easier to express the labor-hours required in terms of 100 ft of material being installed rather than per foot of material. This frequently makes multiplying easier to do. Labor-hour allowances are then generally whole numbers rather than fractions, and this permits the estimator to express the labor-hour requirement more realistically than might be the case if the installation were considered on a per-foot basis.

You should have a good reference table from which you can select labor-hours per 100 ft for various runs of different installation materials. The table should list the differing mixtures of conduit, wire, and tubing (EMT, rigid, polyethylene, and copper) and offer units of labor for each based on either exposed or concealed work. The ceiling spaces of buildings can vary widely in terms of structure, convenient supports for hanging conduit and tubing (which means you don't have to install your own supports), and height. Where your work will be concealed, some liberties can be taken, not in the integrity of the work, but in the appearance. Conduit, tubing, and bundled tubing should be run properly in all situations, but where materials are exposed and where they are to remain visible from the floor, your work should contribute to the acceptability of overhead conditions and not detract from them. Therefore, exposed work should be carefully planned and executed so that it reflects favorably on the contractor who installed it.

Consider a precast concrete ceiling of a gymnasium in a high school. Ductwork, electrical work, and pneumatic work as well as plumbing, rain leaders, and sprinkler piping will all be visible. All visible material will probably be painted, but all the work should be coordinated and executed in a manner that reflects true craftsmanship. When there is a slight offset in the precast construction, attractive offsets should be made in the piping. Proper spacing of anchoring devices, based on careful measurement of the spaces between the runs, is important to the appearance of the work too. And where multiple pipes, conduits, or bundles of tubes turn a corner together in parallel, detailed attention to racking, spacing, and forming is necessary.

In doing work that will be concealed, you should be no less conscientious about making sure your work is performed with integrity than you are in doing work that will be visible. You must make sure that your work meets the necessary codes and is safe and workable. However, attention to making it pretty in appearance is not quite as necessary as for visible work. Spacing of supports for the sake of appearance is not important; they can be staggered where necessary and more convenient. And it is not as important for minor obstructions that saddles be carefully formed in the piping. Suffice it to say that concealed work provides more latitude in terms of installation options than does exposed work, and therefore the labor-hour allowance for a run of concealed piping—whether electric or pneumatic—will be different from the labor-hour allowance for a run of exposed piping.

ASSIGNING LABOR-HOURS FOR SPACE ROUGH-IN

There is no substitute for your own estimates of labor-hours which you have developed from your experience and performance data on past jobs. You are in the best position to judge how much time it will take your crew to perform rough-in work through the spaces of a building. As an experienced estimator, you have a feel for what your crew can do, and you probably have done the work yourself. Therefore, you need to consider carefully what the labor-hours should be for your own shop, basing your decision on what you know. Your techniques, your crew's expertise, and your ability to plan a job and motivate your crew are key ingredients in your decisions about what your shop can accomplish.

Table 9-1 is offered as a general guide to estimating rough-in hours. The figures reflect what I think are national averages among installation personnel, but they may be on the conservative side. If you find that your shop has the ability to improve upon them, you can reconstruct the labor-hour units to more closely fit your shop. You can build in extra time for high work by multiplying the total rough-in hours by 1.3 before you transfer the hours to line 9 of the bid summary sheet (Form 14-1, Chap. 14). Your selection of labor units for rough-in work must be as closely related as possible to your shop's actual production in order for you to be at your most competitive. Therefore, you must form an opinion of the labor units listed in Table 9-1 as quickly as possible so that you can alter them as necessary to fit your situation.

Tubing

Table 9-1 specifies labor-hours for even numbers of lines, beginning with two lines. For a single copper pneumatic line, merely divide in half the time allocated for two lines. Because of its bulk, bundled tube with more than 12 lines should consist of multiple bundles.

Installations of single-line hard-drawn copper in multiple lines are not generally seen; the cost advantages of polyethylene in EMT and bundled tubing are now widely recognized, and this is the preferred approach. Before the widespread use of polyethylene tubing, multiple single-line copper installations were more commonplace.

Setscrew connectors can usually be used in pneumatic raceway systems. The electric raceways in Table 9-1 use compression connectors; although setscrew connectors are cheaper, they are often unacceptable in electrical work. Pneumatic lines in rigid conduit are so rarely used that no labor-hours are given in the table.

Copper pneumatic lines are usually made of hard-drawn copper where they are exposed and soft-drawn copper where they are concealed, and so labor-hours for concealed runs of copper are considerably less than for exposed runs. The time required to install single-line polyethylene tubing is the same as the time required to install a two-tube polyethylene bundle, and so labor-hours for single lines are not shown. The point here is that a two-tube 1/4-in polyethylene bundle is just as easy to run as a single 1/4-in line, and so the labor-hours per 100 ft listed for a two-tube bundle also can be used for single-line polyethylene tubing. For more than two tubes of bare polyethylene, bundling should be used for best efficiency.

Conduit and Wire

In regard to the labor-hours for conduit and wire shown in Table 9-1, derating is not applicable to the limited-energy circuits approved for use in control wiring. EMT sizes for wires are 1/2 in for 2 to 12 conductors, 3/4 in for 14 to 24 conductors, and 1 in for 26 and 28 conductors, based on no. 14 THHN/THWN stranded.

EMT sizes for polyethylene 1/4-in-OD lines (instrument-grade type FR polyethylene tubing) are 3/4 in for 2 to 4 lines, 1 in for up to 6 lines, 1 1/4 in for 8 to 12 lines, 1 1/2 in for up to 16 lines, 2 in for 18 to 26 lines, and 2 1/2 in for 28 to 38 lines.

The EMT sizes for wire given above are taken from published wire fill tables, and the EMT sizes for pneumatic tubing are taken from Table 9-2, which allows for 50 percent fill. Each 1/4-in-OD pneumatic line is considered to occupy a 0.0625-in^2 cross section of space within the raceway.

Table 9-2 • Number of Tubes in Conduit

Trade Size, EMT, in	Area, in^2	No. of 1/4-in Tubes
¾	0.53	4
1	0.86	6
1 ¼	1.5	12
1 ½	2.0	16
2	3.36	26
2 ½	4.79	38

While it might be possible to pull into any size of EMT listed in Table 9-2 more tubes than are indicated in the table, keep in mind that the runs are about 100 ft long and there may be as many as three 90° bends in each run. When estimating a job for a new building, you cannot foresee all the necessary offsets or turns that will be created as a result of structural obstructions or obstructions created by workers in other trades. Ductwork and piping installed by others frequently mean extra work for the installers of control conduits. By estimating conduit that is large enough, you will have sufficient latitude to select raceways that are large enough to minimize labor-hour consumption.

Space Rough-in Guidelines

The following are the important points to consider with reference to Table 9-1:

1. It is unlikely that control estimators will encounter many installations that require exposed runs of hard-drawn single copper lines. Nowadays, multiple

lines of pneumatic control systems in exposed areas, such as equipment rooms, are usually run as polyethylene in EMT conduit. As many as 20 lines often converge into a common run as the control panels in equipment rooms are approached. The consulting engineer who specifies that only single-line hard-drawn copper is allowable in an exposed area is creating extra expense for the project, and would not dictate that requirement unless there was a good reason for it. While hard-drawn copper lines are rarely used for commercial work, they are sometimes called for in industrial work. There are situations where commercial pneumatic temperature controls are used to control environmental conditions in the administrative areas of industrial installations. It is certainly conceivable that the air handlers and other mechanical equipment serving the administrative areas could be located on mezzanines in the industrial spaces. Where the industrial conditions might be harmful to polyethylene, copper lines might be specified because they need to reach the controls apparatus on the air handlers. Thus, labor-hours for installing significant numbers of single-line copper tubes in exposed areas are shown in Table 9-1. Twenty-five years ago, rows of 1/4-in hard-drawn copper lines running through equipment spaces on racks were used in many installations; the lines were works of art and reflected the real craftsmanship of the installers. But the cost of labor and materials and the competitive environment have forced the adoption of newer materials and methods, and polyethylene has taken its rightful place in the world of temperature controls.

2. Obviously, smaller lines, both electric and pneumatic, are sometimes used. Consider the use of no. 16- and no. 18-gauge wire for low-voltage control work and the use of 5/32-in-OD polyethylene tubing and 3/16-in-OD copper tubing. Consult published wire fill tables for smaller electric conductors, and use your good judgment in reducing the number of labor-hours for any 100-ft run. It has been my experience that unless exceptionally large quantities of wire are required, the electrical subcontractors who do control wiring will generally use no. 14 THHN/THWN because of its greater pulling strength and its increased resistance to fracture when joining, striping, and terminating. Additionally, many electrical specifications for construction projects state minimum wire size. No. 12 is usually stipulated as the smallest size permitted for branch circuit power wiring, and no. 14 is frequently listed as the smallest size permitted for control wiring.

In pneumatic installations, 3/16-in-OD copper is sometimes used when an all-copper spec does not permit the use of polyethylene above ceilings and when only a small-size line is needed for the many branch lines above ceilings to devices on terminal units. In this case, 3/16-in soft-drawn copper is used in a single-line configuration, and I suggest that you use the same labor-hours that you would use for single-line polyethylene tubing. Although the connections at each end of a single branch line run take longer to install, soft-drawn copper is more self-supporting than polyethylene and as a result less time is needed to tie off and provide support. Although the run time required for installing 3/16-in-OD soft-drawn copper above ceilings is less than that for installing polyethylene tubing (extra run time is needed to provide adequate support for the polyethylene tubing), the connection time required for copper is greater than that for polyethylene, and thus there is little difference between the total time required for each. With 3/16-in single copper branch line, labor-hours for two-tube polyethylene bundle concealed would be used, or 2 h per 100 ft. (Remember that it takes no longer to run a little two-tube polyethylene bundle than it does to run a single line.) Hard-drawn single lines below a 1/4-in size just aren't seen.

3. Rigid conduit run in concealed spaces is rarely used in commercial temperature control work. However, when electric commercial-type temperature controls are employed in hazardous locations, or are used where the control wiring must be run in concrete slabs, or are called for in power plant projects, then concealed runs of control wiring in rigid conduit are often used. If you look at Table 9-1, you'll see that a little less time is allotted for concealed rigid conduit than for exposed rigid conduit, but not much less. Running rigid conduit is painstaking, important work whether the installation is concealed or exposed.

4. In Table 1, the labor-hours shown for concealed bare copper lines are much less than for exposed bare copper. This is because of the expectation that multiple soft-drawn copper lines will be used, all strapped together, since most specifications routinely allow soft copper lines to be used in place of hard copper ones above ceiling spaces and wherever else they will be concealed. Thus, fewer fittings are required, bending and forming are easier, and the tubing is coiled for greater ease of feeding into ceiling spaces. The cut-in time required for tees is what keeps copper from comparing more favorably with polyethylene from a labor-hours standpoint.

5. Finally, it goes without saying that you must have a tubing layout, rough though it may be, in order to use Table 9-1. When you're first figuring a controls job, you prepare schematics and construct on paper the system you believe will meet the requirements in the spec. With your schematics in hand, overlay the floor plans and determine where lines need to go, and how many. Make use of common raceways for lines going in the same direction. Once done, your roadmap to installation permits ready use of Table 9-1. And your layout will prove very helpful when your bid becomes your job. The success of the job will depend on the estimate. Take the time to develop at least a rudimentary knowledge of numbers and distances of lines, and practice bundling and using common raceways for cost efficiencies.

10

Estimating System Control Panels

Today, control panels are being used more than ever in temperature control installations because of the advantages of centralized readouts and easy control point adjustments inherent in systems that use space and unit-mounted sensors to report to the controllers in the control panels. Another advantage is that the control panel door can be locked to prevent unauthorized system adjustments. The work involved in putting the control devices in the panel and connecting and adjusting the set points needs to be accounted for and entered onto the bid summary sheet (Form 14-1, Chap. 14). The costs for the control panels can be handled in several ways. Listed below are the choices available for building of the control panels and allowing for their costs.

1. Build them in the shop
2. Build them on the job
3. Have them built by others

THE CONTROL PANEL ESTIMATE SHEET

Any of these methods can be entered on Form 10-1, a sample control panel estimate sheet, and you may decide to use any or all of the methods. On Form 10-1, space is provided for the name of the job, the date of the estimate, the name of the estimator, and the page number of this sheet in the total estimate sheets. Space is also provided for the name of the area or room where the panel is located or, in the case of a very large equipment room, the name of the panel as it relates to the pieces of equipment it will serve. For example, a temperature control panel that will contain the controls for air-handling unit 1 and air-handling unit 2 might be labeled "TCP/AHU 1 and 2." This will tell you at a glance that the panel accounted for on this particular estimate sheet is the temperature control panel for air-handling units 1 and 2.

Control Panel Estimate Sheet

Job name ______________________________ Date ____________

Estimator ______________________________ Page _____ of _____

Area or room where panel is located ______________________________

Panel devices	Qty.	Cost/ea., $	Total Cost, $	Labor/ea., h	Total labor, h	Pneu.	Elect.
Subpanels				0.5		—	—
Doors				—	—	—	—
Cutouts		—	—	0.25		—	—
Tubing				—	—	—	—
Wire				—	—	—	—
Total		//////		//////			

Transfer total of panel-mounted device costs (**including** subpanel, door, wire, and tube) to bid summary sheet, line 3

Enter total panel fabrication time in hours on bid summary sheet, line 12*

For special panels fabrication costs, enter quotes on bid summary sheet, line 8

*For in-shop assembly of panels, hours may be factored by 0.8

Total panel hours __________ X 0.8 = __________

Form 10-1 Control panel estimate sheet.

In the first column of Form 10-1, the devices which are to be mounted and connected within the control panel are listed. Here again, complete model numbers are preferred in order to (1) facilitate easy takeoff for ordering, should you be the successful bidder, and (2) ensure that the device selected is the right one for the job. As with any estimate sheet, the quantity is entered and then the price per unit. The total cost for each device is entered in the "Total Cost" column. Labor-hours for mounting, piping and/or wiring, and commissioning the device are listed next, and then the total number of labor-hours are entered in the "Total Labor" column. The quantity required is also entered in either the "Pneu." column or the "Elect." column. These last two columns provide a handy reference whose totals tell you at a glance how many of each type of device are in the panel, and, as well provides a handy check against what you entered originally on your sketches when you took the job off.

There are five preprinted items on Form 10-1: "Subpanels," "Doors," "Cutouts," "Tubing," and "Wire." These items are preprinted on the sheet because some will always be needed on any panel job, and it is all too easy to overlook them when you are concentrating on making sure that the control devices themselves are not overlooked.

Subpanels and Doors

The subpanel, which is the removable back portion of the panel enclosure on which the devices are mounted, is installed after the wiring and tubing are pulled into the enclosure and are ready for terminations. The controls are frequently mounted on the subpanel in the shop on the assembly bench before the subpanel is installed into the panel at the jobsite. If instead the controls are mounted in the field, the subpanel is installed after rough-in to the enclosure is done, and then the controls are field-mounted. Note that Form 10-1 makes provision for multiplying control panel unit labor-hours by a factor of 0.8 when the work is performed in the shop rather than in the field. In-shop conditions are usually more conducive to greater productivity than jobsite conditions, and so the panel labor-hour units in Table 10-1, which were calculated for work performed in the field, may be factored when this work is done in the shop.

Since doors must always be ordered for the enclosure, they must be accounted for on every panel estimate, and consequently an entry for doors is also preprinted on Form 10-1.

You should investigate possible sources for the whole control panel assembly, which includes the door. Perhaps a local fabricator would be willing to give you a standard price on a particular style of panel, or perhaps you might find a large fabricator who has one style that you can use with some degree of regularity and who might be willing to give you a break on lot shipments.

Cutouts

Invariably, the controls contractor is faced with the need to cut out holes for gauges, switches, or recorders in the door of the panel. Whereas 1/2 h is allowed for the actual placement of the subpanel within the enclosure (once it's on the job), 1/4 h is allowed for each cutout. These time allowances are also preprinted on Form 10-1. The 1/4 h for cutouts covers regular cutouts which are round, and which can be made with hydraulically operated punches with portable hand pumps. For square or rectangular openings, you will have to consider the difficulty involved and the procedure you will follow and then make appropriate labor-hour allowances for the various special openings you might require. If you decide to have others make the cutouts, treat the costs as special fabrication costs and enter them on line 8 of the bid summary sheet (Chap. 14, Form 14-1). Portions of the panel under consideration might be constructed by others, or the whole unit might be assembled by others. In either case, list the devices on the panel estimate sheet, just as you would if you were building the panel, because the responsibility for designing the controls application is still yours, and you will probably be furnishing the controls to go in the panel, or in the door as the case may be, for others to mount. List the quantities and the costs for the devices, as before, but enter the labor-hours for mounting and piping or wiring the devices, which will now be done outside your own shop, as special fabrication costs on line 8 of the bid summary sheet (Form 14-1). You can see, then, that some latitude can be exercised in the use of the panel estimate sheet, where part or all of the costs of the panel construction may be delegated to a fabricator through your purchasing system. It is these costs that must be carried over to line 8 of the bid summary sheet.

Tubing and Wire

The last two items preprinted on the panel estimate form are tubing and wire. Depending on whether the control selected is pneumatic or electric, one or the other or perhaps both will be required. Just estimate roughly an allowance of a few feet of tube or wire or both, as the case may be, and show the total footages for the panel being estimated. Assign a price and extend the costs so that they become a part of the panel costs. Total the columns and transfer the totals to their appropriate places on the bid summary sheet.

ASSIGNING LABOR-HOURS

Table 10-1 lists labor-hours for various devices likely to be found in temperature control panels. Remember, they are jobsite figures, subject to factoring by 0.8 when the work can be performed in the shop on the bench. The labor-hours in the table reflect installation, connection, and setup time. Note that where a dash is entered, labor-hours are not applicable. For example, because nowadays most control panels in commercial temperature control work are actually enclosed cabinets, it is highly unlikely that you would mount a relay or a contactor on the door. The same holds true for the other devices after which a dash appears in the "Door-Mounted" column. These devices are usually mounted inside the control panel.

Dashes are also used in Table 10-1 where the device listed is generically pneumatic rather than electric, or vice versa. Contactors are electric, not pneumatic. Air headers are pneumatic, not electric. Terminal strips are electric, although a pneumatic terminal strip with numbered through-ported openings is available too. The time allotted for terminal strips is low because the connections are allowed for in the devices listed on Form 10-1. PEs and EPs require

Table 10-1 • Temperature Control Panel Labor-Hours

Panel Devices	Door-Mounted Pneu.	Door-Mounted Elect.	Mounted Inside Pneu.	Mounted Inside Elect.
Relays	— —*	— —	0.5	0.75
Contactors	— —	— —	— —	1.0
Switches	0.75	1.0	0.5	0.75
Controllers				
Single input	— —	— —	1.2	1.8
Dual input	— —	— —	1.75	2.3
Triple input	— —	— —	2.3	2.5
PEs	— —	— —	0.5†	0.5†
EPs	— —	— —	0.5†	0.5†
Time clocks	— —	— —	1.0	1.0‡
Gauges	0.75	1.0	0.5	0.75
Dial thermometers	1.25	— —	1.0	— —
Air headers§	— —	— —	0.5	— —
Terminal strips§	— —	— —	0.5	0.5
Shutoff valves§	— —	— —	0.25	— —
Fuse blocks§	— —	— —	— —	0.25
PRVs§	— —	— —	0.5	— —
Transformers§	— —	— —	— —	0.75
Receptacles§	— —	— —	— —	0.5
Power supplies	— —	— —	— —	0.75
Recorders	1.25	1.25	— —	— —
Indicating recorders	1.5	1.5	— —	— —
Jacks§	0.25	0.25	0.25	0.25

* Labor-hours are not applicable.

† Add the hours for pneumatic and electric installation together; the device requires both.

‡ Energy-optimizer time clocks may also need a pneumatic signal.

§ Identify and enter in "Total Cost" column on Form 10-1 if not already accounted for.

both tube and wire, and so the 1/2 h in both the pneumatic and electric columns must be added together.

In regard to time clocks, Table 10-1 reminds you that optimizing energy-saving time clocks that use a building's programmed U factor and outside air temperatures frequently utilize a pneumatic signal from an outdoor air temperature sensor. When this is the case, the 1 h shown for time clocks in the "Pneu." column must be added to the 1 h shown in the "Elect." column. Otherwise, the pneumatic hour is not counted. Electronic time clocks are assigned 1 h because programming them is usually quite simple once you are familiar with how it's done. When transformers are required for low-voltage power to electronic time clocks, the labor-hour allowance from the transformer line is added.

Table 10-1 also reminds you to make sure that several items listed in the table must be listed on Form 10-1 as well. The reminder is just a memory jogger to prevent their being overlooked. For example, a receptacle might be useful for plugging in tools and trouble lights that can be used for field checking and troubleshooting later on if the panel is in an out-of-the-way area. If so, list the item on Form 10-1 and price it in with the other devices.

The dial thermometers listed in the table are assigned labor-hours only in the pneumatic column, because they are the capillary type, necessitating the installation of the capillary from the control panel over to the unit where the measurement is to be taken. Pneumatic dial thermometers should be assigned the same labor-hours as the gauges. The stringing of the capillary is obviously a field installation item and cannot be factored by 0.8 when the panel is shop-fabricated; however, unless there are a significant number of them, the slight error is probably acceptable. When a significant number of panel-mounted capillary-type thermometers are involved, don't factor them by 0.8 if the panel is shop-fabricated. Instead, circle the labor-hours on the panel estimate form and omit them from the factoring. Factor the remaining panel labor-hours and then add the number of capillary thermometer labor-hours to the factored total. The labor-hours for stringing capillaries should appear on Form 3-1.

The first four items in Table 10-1—relays, contactors, switches, and controllers (other than indicating or recording-type controllers)—are each assigned average times for their installation even though variations in characteristics of each type are common. Usually various types of each device are found in a typical control panel, and so the allowances are more realistic than might be imagined. As with all estimating, the special requirements associated with any job must be identified and considered independently of these labor-hour allowances, and so you must use your own good judgment in making the estimating methods discussed in this book of value to you. Good record keeping is a tremendous help. Where possible, computerize your data, and from it develop factors that can assist you in making decisions on jobs that fall into distinct categories. Even better, also computerize your completed jobs by geographical location, to allow for varying labor rates and differing work attitudes that can influence the efficiency of any job.

CONTROL PANEL COST AND CONSTRUCTION

National manufacturers often construct temperature control panels in centralized production facilities from drawings prepared in a branch office. The branch office is charged for the completed panel with the control components in place. The panels may or may not be piped, but usually they are wired to a terminal strip. Panel costs frequently were a concern to branches in very competitive environments, and as a result of their need to be more competitive, many branches would assemble their own panels using enclosures from conventional suppliers. When this happened, the central assembly facilities of large corporations would sustain a drop in volume of shipped panels, and generally cost-cutting measures would be studied by the home offices to lower the cost of assembled panels to the branches. The point here is that the branch offices of large corporations are frequently in no better competitive posture than the independent controls installation office when it comes to assembled panel costs. Controls contractors have similar options in regard to control panels, because they can assemble them in their shop or on the job or can have someone else do it for them altogether. Therefore, use your expertise and ingenuity in figuring panel costs. Shop for the best deal on enclosures and try to obtain quantity price breaks from your supplier. Design your own enclosure, if you like. Control panel costs are an item which can give you an edge over some of your national competitors. And when large quantities of panels are involved, in-house assembly-line techniques are sure to provide cost savings.

Control Panel Rings, Subpanels, and Doors

The three-piece control panel continues to be the most popular choice because of the flexibility it allows during installation. The three pieces are the panel ring, the removable subpanel, and the removable door. As mentioned before, the ring is mounted early in the installation to provide a point toward which to work with conduits and tubing. The word "ring" is used, not because the panel ring is round, but because it is open in the front and back; neither the door nor the subpanel is in place when the ring is first mounted. After rough-in work has brought the tubes and wiring to the ring and they are hanging into the panel awaiting termination, then the subpanel is installed and the tubes and wires properly connected. Then, the door is attached, and the panel is complete.

You also need to consider whether or not you want the subpanel to be prepunched. Self-tapping screws are very handy, and normally the thickness of the subpanel metal does not prohibit their use. Prepunching might be an item of cost you can do without. Prefabricated subpanels that are perforated are also available; drilling is not necessary, and controls can be mounted more quickly to the subpanel.

Figure 10-1 depicts a typical panel ring and highlights some of its important features. The panel ring (A) is 48 × 36 in. The figure shows the formed jam (B) for the removable door and highlights the raised lip (C) in back, which

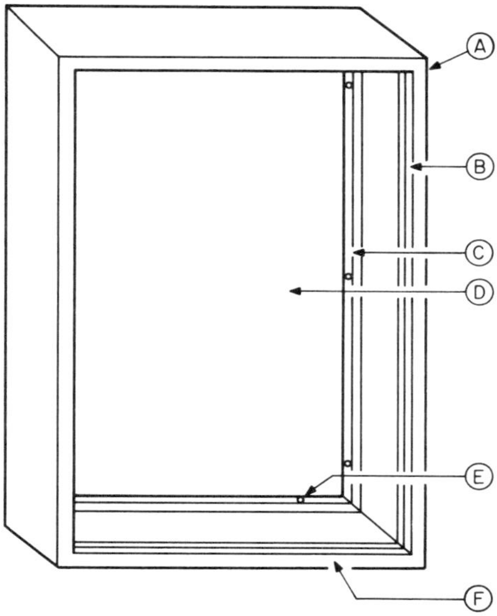

Figure 10-1 Typical control panel ring. A, typical 48 × 36 in ring; B, formed jam for removable door; C, raised lip in back that creates a 1-in standoff for subpanel; D, back, open to the wall until subpanel set; E, prepunched holes for hanging rings; F, personalization of panels with firm's identifying color.

creates a 1-in standoff for the subpanel. The back (D) is open to the wall until the subpanel is set. Holes have been prepunched for hanging the ring. Note that you can personalize the panels (F) by using the same color (of your choice) on all your shop's panels—this way the color can be readily identified with your firm.

Control panel rings fabricated from code-gauge steel are given a rating of NEMA 1 by the National Electrical Manufacturers Association (NEMA) and are suitable for general-purpose indoor applications. Practically all commercial mechanical equipment rooms can be served with this type of enclosure, since special enclosure requirements in these general areas are rare.

Sizes of control panel rings vary among manufacturers, and often size is dictated by actual jobsite requirements. Although probably the size most often used is 48 in high by 36 in wide by 12 in deep (shown in Fig. 10-1), there is no reason why the basic design couldn't be produced in two or three readily available or readily stocked sizes to accommodate varying quantities of controls. The removable door should be designed so that it can be opened from either the left or the right. This feature is important when wall space in the equipment room is at a premium and fitting in the equipment becomes difficult. Since the screws used to mount the controls on the subpanel go clear through the subpanel, clearance behind the subpanel needs to be allowed for the tips of the screws. The mounting lip raises the subpanel just about an inch, thus providing enough clear space to make installation of screws easy.

When hung initially, the ring is open in the back and the front. Bolts or screws for mounting the ring are installed through the gutter area between the sides of the ring and the raised lip onto which the subpanel is later mounted. The gutter area can lie hard to the wall; frequently it is prepunched, and if not, it can be easily drilled in the field. The open back, which is later covered by the subpanel, makes mounting of the panel on uneven surfaces easier, as the 1-in clearance can accommodate minor uneven surfaces nicely. When the lip is punched, the subpanel merely has to be screwed onto it. The lip is rolled in far enough so that its outside dimensions are smaller than the opening for the door.

Generally the panel ring and the door are painted the same color. The ring should be painted all around, inside and out. The door needs to be painted only on the outside. The subpanel does not require paint, but should be made of galvanized metal or perhaps aluminum. You might decide to paint all your panels, regardless of size, with a special identifying color, which would serve to advertise the presence of your company's crew and materials on the job.

Knockouts

Knockouts are desirable to have. Multiple concentric knockouts are ideal, and should be located on all four sides at each corner, in sizes up to at least 2 in; the smallest portion of a multiple concentric knockout should be 3/4 in. Additionally, 1/2-in and 3/4-in multiple concentric knockouts should be placed in the center at the top and bottom, preferably in a row of four. For multipanel installations where several knockouts need to be grouped side by side or in a combination of side by side and over-under, the knockouts should line up, permitting the use of chase nipples installed through abutting openings in the panels. Then, interpanel wiring and tubing can be used without the necessity of using conduit nipples. The inset feature of the door opening ensures that the doors can be opened and removed even when the panels are side by side in close proximity.

FREESTANDING INSTALLATIONS

When desired, the panels can be bolted together and mounted on fabricated legs for a freestanding installation, or braced off a wall or the side of an air handler. When wall space is not available, the freestanding approach offers options in placement of panels that frequently work to the benefit of the controls contractor because the panels wind up being closer to the unit involved. For example, consider an air handler in the middle of a fan room. If the panel is located on a wall, the tubing and wire must travel from the controls on the unit overhead to the panel on the wall. If a freestanding panel is installed closer to the air handler, runs of tubing and wire are reduced, resulting in savings in time and materials.

MOUNTING CONTROL PANELS ON EQUIPMENT

When locating panels near the equipment they serve, take care not to block access openings in the equipment they are to be mounted near. When possible, rather than using a freestanding approach, mount the panel *on* the unit. This will save even more material and labor. A panel can be mounted on a unit as long as there is a side that cannot be removed and as long as that side does not provide access to the equipment. Then you can be sure that the panel won't block future efforts to get into the unit.

11

Control System Time Framing and Labor Planning

Job management is particularly important in temperature control work. In making decisions regarding deployment of your workers, you need to consider the penalties of staffing the job too early as well as the repercussions of waiting too long. Sending workers to a jobsite before the site is ready for them can be costly in downtime, because your workers will have to wait for various segments to be completed before they can perform their jobs. On the other hand, not sending your workers to a jobsite until the optimum time has passed can cause your customer to become concerned as jobsite supervisory personnel begin to worry that workers in other trades will cover up areas where your controls are to be installed and you can no longer rough-in with any degree of efficiency. Ideally, your appraisal of the job at the bidding stage, coupled with input from those to whom you are submitting the bid, will allow you to estimate with reasonable accuracy the time frame during which your installation activities must take place. By examining the hourly rates that are forecast for the period during which the job will be in progress, you will have a good idea of the hourly rates that are most likely to be in effect at the time the work is performed. Pay careful attention to those periods when rough-in and fan room work will be accomplished, because these two areas represent the bulk of your labor-hours. The more accurately you can gauge the time when this work will be done, the more realistic your estimate will be.

ESTABLISHING A TIME FRAME

In looking at the overall hours estimated for the job, you can see that only effective production time has been estimated. As an estimator, you have planned on labor-hour expenditures as mandated by the structure of your control system and by the difficulties of installation as viewed from the contract drawings. Your estimate is tight. There is no room for unnecessary jobsite visits. Now look at the completion date called for in the bid set, and briefly lay out a time frame upon which you can mentally superimpose your labor-hour allocation. This will help establish the year and months of the control system construction activity and subsequently the hourly rates for labor.

Form 11-1, a labor graph for time-framing controls work, can help in pinpointing your involvement time. The graph consists of a diagonal line traversing a time-and-percentage grid. The diagonal line helps plot where labor will be needed. The bottom of the line represents the time your workers will begin the job, and the top of the line represents the projected time that the general contractor will turn the job over to the owner. Be careful to think in terms of the time that your workers will begin the job, and not the time that the job itself will begin. The presence of your crew on the job may not be required until the roof is on and the installation of ductwork has begun. This is the beginning point of the temperature control work in a conventional installation, where the work will be predominantly in the ceiling space.

Not only will you most likely be the project estimator, but the project will be yours when you win your bid for the job. As a project manager, you can make or break your job by successful or unsuccessful labor management. You'll find that there are no spare hours in your estimate for installation of a control system. It makes sense then that the estimated hours be expended at the optimum time. Therefore, *you* must run the job. You must make your

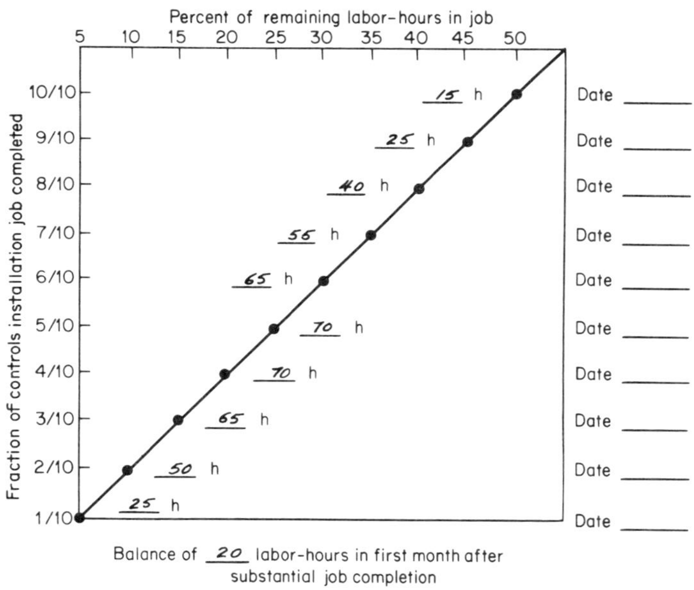

Form 11-1 Controls work time-framing labor graph. Example plotted on graph is a control job that takes 500 h. If total construction time for the project is 18 months, control work begins at the ninth month into construction.

own decisions about labor requirements, and you must visit each job to know where it stands and what it requires. If you rely on hourly paid supervisors to tell you what the job needs, your best interests will not be met. If you are to succeed, you must be knowledgeable about installation work, and you must continually update your information so that you can judge the proper time during the construction work to install your temperature control system. If you begin the job too early, your work will get damaged by workers in other trades. Remember, pneumatic tubing is predominately 1/4-in copper or polyethylene, and if it gets in the way of heavier construction materials, the results can be costly. You decide when to place workers on the job, and be sure they have a riser diagram and/or a floor plan layout showing the tubing and how it is to be run. At the time of rough-in, schematic control diagrams are unnecessary for the rough-in installers. Instead, they should have accurate tubing routing diagrams in floor-plan view, showing materials and sizes. As unpleasant as adjustments in your hourly worker force can be, you must reduce your installation force, just as you must increase it, to fit the workload.

Form 11-1 is designed to bring up your installation activity when it is time for you to begin meaningful work on the job. Your ability to use the graph successfully, therefore, improves proportionately with your experience and with your familiarity with typical construction progress on the type of work you're estimating. In Form 11-1, maximum labor-force output falls within the third tenth and the seventh tenth of your predicted time frame. Remember that this is probably 50 to 100 percent of the total time span of the overall job as viewed by the general contractor. Thus it is important that construction is at the point you want it to be at when you start your installation activity. Consequently, your accurate estimate of the proper time to deploy your crew increases the odds of labor rates being at the level you expect when your jobsite work commences.

In the graph, the percentages of remaining labor-hours are listed at the top in ascending order from left to right. At the left-hand side of the graph, the time frame of control construction work is divided into tenths in ascending order from bottom to top. The spaces provided on the right permit you to enter the dates that your installation activity reaches each level. The diagonal line from the lower left to the upper right represents your jobsite activity curve. As the diagonal reaches each junction of tenths and percentages, you enter the labor-hours, calculated by multiplying the remaining job labor-hours by the appropriate percentage.

To illustrate, numbers have been entered for a 500-h job. The first entry is 25 h, because 5 percent of 500 h represents the first calculated entry. At the second tenth of job progress, 10 percent of 475 h is realized, rounded, and entered at the junction of 10 percent and the second tenth. By the time the third tenth of job progress is reached, 425 h remain in the job, and 15 percent is rounded to give the 65-h entry. Four-tenths of the way through the estimated hourly labor, 360 h remain, and the 20 percent factor yields 72, which is rounded to 70 hs. The process continues until at ten-tenths of the way through, the final entry of 15 h appears. And this should put you out of the building, along with the general contractor, as far as installation is concerned.

The hours that have been plotted add to a total of 480, leaving a balance of 20. These final 20 h are for tying up loose ends in the first month after the job has been substantially completed. Tagging devices, tightening damper linkages, calibrating space thermostats, and making other adjustments as may become evident and/or appear on the punch list are the kinds of jobs you would expect to do at that time. It is thus realistic to plan for a portion of the installers' time to be spent as the owner takes over.

The graduating percentage method for plotting the labor-hour expenditure is offered as a tool for forecasting at bid time how the hours will fall in relation to specific calendar months. It reflects the usage of labor on an average commercial job, and should be sufficiently accurate to permit bid-time planning of future labor rates. Your must provide the start date and completion date and decide what the hourly rate will most likely be.

LABOR PLANNING

With the time frame of the project established, you can direct your attention toward labor planning. From Form 11-1, sufficient information will have been developed to give you a feel for when the job will break, and you will have selected your hourly rates according to the periods of time in which your installation activities will most likely occur. Labor planning is essential to forecasting increases or reductions in your workforce, and its input at bid time helps sales management personnel judge the urgency of doing a particular job in-house. Some rudimentary involvement with your production people is advisable at bid time, for reasons you will see a little later in this chapter. The objective for the moment is to gain some familiarity with the labor plan and to examine its construction. Job estimators should be comfortable with the application of the labor plan and should recognize the ingredients of the work it covers.

THE FIVE BASIC CATEGORIES OF CONTROL SYSTEM ESTIMATING

Control system estimating can be divided into five basic categories, each of which forms a specific segment of work with its particular scope of activity. Space rough-in is the first and consists of installing tubing or conduit and wire throughout a building, usually above the ceiling, for purposes of interconnecting room temperature devices with fan room devices so that space temperature requirements can be transmitted back to the controls at the air-conditioning equipment. The second category is equipment room piping or wiring, where control devices are installed on air handlers and other equipment and are tied into temperature controllers, motor starters, valve and damper actuators, and the like. The third category deals with control panel construction; this can be conducted simultaneously with space rough-in or fan room piping or wiring, because quite frequently control panels are fabricated in the shop on a bench and then inserted into field enclosures later on. Mounting of control panel enclosures is usually a part of equipment room work (category 2), and tubes and wires are pulled in for later termination to devices mounted on subpanels in the shop. The fourth category is mounting of space controls, such as room thermostats. There can be numerous thermostats in large buildings, and mounting is usually reserved until after painting is finished. Lastly, system checkout and setup are performed—each piece of control equipment is checked to verify that it is operating properly, and set points are assigned.

Study these five basic areas of control system estimating for a moment.

1. Space rough-in
2. Equipment room piping or wiring
3. Control panel construction
4. Mounting of space controls
5. System checkout and setup

The estimating system presented in this book is based on these five areas of work, and thus the estimating procedure takes into account the elements of installation for these five areas.

Estimate Sheets

The estimating procedure is based on four sheets: the equipment room labor and materials estimate sheet, Form 3-1; the space-rough-in estimate sheet, Form 9-1; the control panel estimate sheet, Form 10-1; and the bid summary sheet, Form 14-1. Mounting of space controls and checkout and setup of space controls are allowed for on Form 9-1, and checkout and setup of equipment room devices are allowed for on Form 9-1 as well. Thus, a mix of talents is often found on Forms 3-1 and 9-1, since rough-in and calibration are frequently handled by two different types of worker but are estimated and summarized on the one sheet pertinent to the area in which the work is domained. Labor classification is an installation supervision function and is not necessarily the concern of the job estimator. However, if you are a project manager in an office structure where one individual estimates the job, bids it, books it, oversees its engineering, and eventually oversees its installation and conclusion, it is advisable to examine briefly the personnel involved in the installation field work in light of the classes of labor you will require. It is likely that this aspect of installation and its relation to estimating will be of benefit only on large projects, where the long-range commitments of the shop's labor force might necessitate some minimal labor planning at bid time. Thus, Form 11-2, a sample labor planning sheet, holds perhaps some value for the estimator during bid preparation, but it is essential to the installation supervisors for labor management once the job is underway. For that reason, it

Labor Planning Sheet

Space rough-in   Equipment room work   Control panel work
Mounting of space controls   Checkout and setup

Date prepared ______________________   Prepared by ______________________

Enter hours required per month on each job and completion date

Job name	Month												Completion
Total monthly hours required													
Total workers required (divide by 176)													

Form 11-2 Control systems installation labor planning sheet.

is appropriate to review Form 11-2 now, for it is extremely helpful in putting to use the benefits of the information from Form 11-1, your time frame for labor costs.

When you look at Form 11-2, it helps to keep in mind the five categories of controls installations as well as the type of work being planned. For this reason, the five categories are noted at the top of each sheet as a reminder that as you plan the labor for the job, you must allow for these various segments of installers' work to occur in their proper order. From the information, you can anticipate shortfalls in expertise for the more technical activities of the job.

Because labor planning is most effective when you look at a new job in light of the backlog of in-house work, each month the construction coordinator should list all jobs on a labor planning sheet and post it where it can be referred to by members of the sales force who are estimating work for future installation. In the ideal situation, the estimating and sales-force data are entered on the monthly labor planning sheets, so that the pending workload of the shop can be considered as estimating takes place. In this way, sales efforts can be increased when valleys in the workload are foreseen, and jobs that are projected to break when the shop needs work can be more aggressively pursued.

As an estimating tool, Form 11-2 might be utilized only on large projects, with the input of the construction personnel included too. In this way, time framing of hourly costs for hourly installers is more easily achieved, as the monthly breakdown of the job is studied with an eye toward the varying amounts of activity shown on Form 11-1.

Job management is really everyone's job, although the thrust of such efforts belongs to the construction supervisors and/or coordinators. Estimators contribute to the management of jobs through sufficient planning and through communication with the installation supervisory people. Sufficient planning ensures that the job can be done for the hours allowed. Communication alerts the supervisory people to the special needs that the sales force should consider in order to land a job and, once the job is won, to the critical path of the project.

The Labor Planning Sheet

In Form 11-2, the five elements of a typical controls installation are shown at the top merely as a reminder that as you prepare the plan, you should consider each aspect in order to ensure as accurate a labor spread as can be prepared. Enter the date on which the plan is made so that all concerned can readily appraise its validity as a current document. List all jobs in-house in the order that they were booked. Next, enter the upcoming 12 months. Ideally, this plan is prepared on the first of each month. After the months have been listed, consider each job in order, and forecast for each month the number of labor-hours that you anticipate the project will require. Needless to say, your work in-house will start and end in varying months. In the column labeled "Completion," show the month and year that you believe the job will be finished.

Once you have done the forecasting, total the monthly labor-hours for each month and then enter each total on the line "Total Monthly Hours Required." This plan is now a projection of your work in-house and of the way you expect to allocate your work force to meet construction completion schedules. Finally, divide each monthly labor-hour total by 176 to derive the number of installation people you will need in any given month. This number becomes the final entry at the bottom of the sheet, on the line labeled "Total Workers Required."

The number 176 represents the average number of labor-hours in a month. To arrive at this number, consider a 30-day month with 4 weekends, and subtract 8 days (for the weekends) from 30, which leaves 22 working days. At 8 h/day, the number of working hours per month is 176 (8 × 22). Months with 31 days will perhaps provide another working day, while February will contain fewer working days. Time for holidays will also have to be subtracted from some months, but overall an average of 176 seems reasonable.

CONTACTS WITH CUSTOMERS

I was an installation supervisor for many years, and I found it was advisable to contact my mechanical contractor customers by phone at the time that I prepared my labor plan. Not only did this give me a chance to say hello and indicate my interest in their job, but also it provided me with additional insight into when the job would be completed, for my customers invariably knew of a glitch or two that might slow the job down. These phone contacts, made on a regular basis, tell your customers that their jobs are important to you, and the calls invite your customers to update your plan with all the information that they have on the job. Additionally, your customers gain from your calls; they can see what your needs are and can help if there are areas in which you are being held up. Feel free to tell your customers that you are in the process of developing your regularly prepared labor plan and that you would appreciate their opinion of what the actual completion date of the project will be. This personal touch will be welcomed and will show your continuing interest in the progress of the job.

12

Knowing Your Electrical Controls Job

As often as not, a job specification calls for the installation, pneumatic or electric, to be the responsibility of the temperature controls contractor. It is not unusual for the specification that governs the control work to include a sentence such as, "All control and interlock wiring required to interconnect and control the HVAC equipment shall be included in this section of the specifications." In itself, this statement is quite broad and is frequently construed to cover control wiring pertinent to all the controls on the job related to the HVAC system, whether furnished by the controls contractor or not.

CLARIFYING RESPONSIBILITY

It is extremely important to be thoroughly familiar with the complete control wiring requirements of a job if you are to undertake all that could be required on the job. Otherwise, you need to clarify your bid as to exactly what wiring you are including. Chiller interlock wiring, mounting and wiring of factory-furnished controls shipped loose with equipment, and motor starter interlocking of pumps and fans, for example, are jobs which can show up later in the project and which you might be expected to do. If you have neglected to recognize this work and either have not included it in your estimate or have not stipulated that you will not do it, you're in for a surprise, and one that will not be too pleasant. When your customer's purchase order says, "Furnish and install a complete control system per plans and specifications" and then specifically calls out the temperature control section, you are tied to interpretations of those plans and specifications, in spite of what your proposal may have said, particularly if you accept the purchase order without exception. In this case, your signed proposal would have been a better instrument for you than the purchase order. Your proposal probably would have clarified your position regarding certain areas of work that the purchase order, with its "per plans and specifications," will hang you for.

Consider the following sample jobs of controls wiring that may not be apparent from merely reading the temperature control section of the job specifications.

- Wiring cooling tower sump heater controls
- Interlocking air-handling-unit filter advance motors
- Tying in fire alarm control panels to air-handling-unit smoke detectors or firestats
- Wiring boiler controls
- Interlocking fuel oil pumps and fuel line solenoid valves
- Wiring emergency shutdown switches
- Wiring fan coil unit controls
- Wiring chilled water flow switches to water chillers
- Wiring louver actuators in exhaust fans and power roof ventilators
- Doing incidental supplementary wiring to effect the sequence described using components furnished by others
- Wiring cooling tower capacity control dampers
- Interconnecting remote condensing units in split systems

In the above list, none of these jobs may have been described in the temperature controls section of the job specifications dealing with the devices to be furnished and installed by the controls contractor, but a catchall phrase that covers "all control wiring" has caught many control companies, and they have had to perform the work. Usually, the sequence of operation and/or the control diagrams pertinent to the work of the controls contractor are sufficient for determining the specific devices required. But for the sample jobs just listed, there may have been no mention in the temperature control specs of the sequence that would have alerted the controls contractor to include this work.

What often happens is that the matter arises on the job when components need to be wired. Your customer, the mechanical contractor, looks to you to perform "all control wiring." The jobsite electrical contractor will not do the wiring because electrical contract work is detailed on the electrical sheets of the project drawings, and further, there may be a phrase in the electrical section of the specification that exempts the electrical contractor from performing HVAC control and interlock wiring.

Let's examine the sample job of control wiring to see how the need for performing such wiring might arise on a job.

Cooling Tower Sump Heaters

In the case of the cooling tower sump heater, the cooling tower was purchased as a package by the mechanical contractor, and it was shipped to the job and set in place. Then the mechanical contractor noticed that a thermostat in a weatherproof enclosure had been shipped loose with the tower. Nothing on the electrical drawings alerted the electrical contractor to the need for its wiring, and the cooling tower specification merely said, "Tower to have thermostatically controlled sump heater for freeze protection."

Now someone has to perform the work of mounting and wiring the thermostat to the sump heater, which is an electric immersion heater, complete with low-level cutoff. On the electrical drawings, a designated circuit for power to the heater probably was shown on the plan view, but it's a safe bet that the control wire to the sump thermostat wasn't shown. And the specs said: "Control wiring by the controls contractor." Guess who inherits the responsibility for this segment of work? Although the work was overlooked by everyone, that inclusive phrase in the controls section can be costly to you unless you have clarified your position that you are not responsible for that kind of work and your agreement with your customer recognizes that clarification. The architect's position will be that a complete installation was called for and that conflicts over who does the work are to be resolved by the general contractor.

Air-Handling-Unit Filters

Many times a packaged air handler features an electric filter drive mechanism activated by a differential pressure switch to automatically drive new, clean filter media into place when the media in position becomes dirty, and to energize a filter media run-out light when the media in the filter roll is all used up and needs to be replaced. In addition, the recommended control wiring diagram inside the control box cover on the filter drive unit probably shows that interlocking with the operation of the air-handling-unit fan is advisable too. At the least, the requirements here will probably be for (1) a 120-V circuit to the media advance motor, (2) extension of the circuit to the differential pressure switch, and (3) inclusion of an auxiliary contact in the motor starter for the fan of the air handler in the circuit. Who will do this wiring? The wiring was not included on the electrical drawings. Your customer (the mechanical contractor) didn't realize that the wiring would need to be done. Did you? Is it covered in your bid?

You should become very familiar with the job, and learn to clarify your position with respect to incidental control wiring of equipment furnished by others. You don't want to leave your customer in the lurch. You need to become familiar with routine circumstances like this and learn to protect yourself and your customer through comprehension and clarification of what's needed.

Fire Alarm Control Panels

Fire alarm control panels are ordinarily specified and detailed in the electrical section of the specifications. And fire alarm risers on the electrical drawings will do a good job of showing the electrical contractor what is required. But one little area is frequently overlooked and is not shown on the fire alarm risers. This area is fire alarm control panel tie-ins to remote safety devices. The temperature control section in the specifications states that "Smoke detectors or firestats furnished by the controls contractor shall stop the air handler should they sense products of combustion or high temperatures indicating the presence of fire." These safety devices are located in the return and/or supply ductwork of the air handler, and they are included in the control specifications, along with the wiring to the motor starter of the air handler. In the electrical section, under the description of the fire alarm system which the electrical contractor is to furnish and install, you may find a sentence that says, "provision shall be made for alarm initiation from fire or smoke devices on the air-handling unit." Such an instruction, coming under the specification for the hardware of the fire alarm system itself, merely tells the fire alarm equipment supplier to provide terminals in the fire alarm control panel for remote wiring to firestats and smoke detectors on the air handler. But who performs the wiring? The fire alarm riser didn't show the wiring. The devices to be tied in are furnished under the temperature control section of the specification. And the architect's interpretation of the spec may be that these terminals were to be present so that the temperature controls contractor could tie into them. Guess who runs the circuit from the duct-mounted safeties to the fire alarm control panel? Didn't your purchase order from your customer say "per plans and specifications"?

Boiler Controls

Boiler controls often are shipped as loose components to be field-piped (in the case of gas line solenoids and safeties) and field-wired. The components can include the boiler combination aquastat and high-limit, low-water cutoff; flame detection control; and, as mentioned, gas line solenoids, in the case of a gas-fired unit. In all likelihood, the electrical drawings show a 120-V circuit feeding the boiler, and that's all. The boiler spec may have detailed the component control devices to be furnished by the boiler manufacturer, but it may not have said "prewired." The mechanical contractor purchased the boiler from the manufacturer's representative according to the interpretation which the rep made at the time of takeoff. It's so easy to forget to check—are the controls that come with the boiler factory-wired into the boiler control box? During the intensity of bid preparation, the mechanical contractor is hard-pressed to check out the intricacies of each supplier's quote. The mechanical contractor counts on each quote to cover what is required. And when the control section of the specifications says that the controls contractor "shall furnish control and interlock wiring," you can be sure that the controls contractor will be expected to handle the requirements for the control wiring of mechanical equipment.

Fuel Oil Pumps and Fuel Line Solenoid Valves

Even when an item associated with the boiler is not mounted on the boiler per se, it can become an item of contention in regard to who is to perform the wiring. Consider fuel oil pumps feeding commercial oil burners and remote entrance-point fuel line solenoid valves. The diagrams pasted on the back of the cover on the control box of the boiler recommend interlocking of the fuel oil pump (when one is required) and extension of the safety shutdown circuit to the entrance-point fuel line solenoid valve. Because of the increased emphasis on safety, electrically actuated fuel line solenoid valves are frequently provided so that when the boiler shuts down for any reason, the fuel line is shut off at the point where it enters the boiler room. Who is to do this wiring? A close look at the mechanical drawings in the contract drawing set would have revealed that in an isometric of the boiler fuel feed system, an electric solenoid valve was shown as part of the fuel line piping. It's a safe bet, though, that the wiring of this fuel line solenoid valve was not shown anywhere in the contract documents. But the control section of the specs said "all control and interlock wiring." Who do you suppose will inherit this work if the architect renders an interpretation of the spec at the request of the general contractor? The work needs to be covered. It doesn't matter who includes it in the estimate, but somebody should. It pays to clarify.

Emergency Shutdown Switches

Along these same lines, panic or emergency shutdown switches are frequently shown located outside the doors to boiler rooms, and often they are shown not on the electrical drawings but on the mechanical drawings. In actuality, a control circuit effects remote shutdown from the boiler room doorway if an emergency arises. You would not expect the electrical contractor to cover this wiring if it is not shown on the electrical drawings and not covered in the electrical section of the specifications. But if the mechanical drawings do show it and if the controls contractor is responsible for control and interlock wiring, who has to do it? Even better, who furnishes the emergency shutdown switch? Did you enter it on your estimate sheet? The point here, which bears repeating over and over, is that if you are going to make a profit at what you do, you must enter into contracts that clearly delineate what you will be doing. Many assume, because a controls contractor is involved on a project, that the controls contractor will intervene in matters pertaining to any controls, no matter what they're on or who furnished them. You must estimate the work if you're expected to perform it. You must clarify your bid if there is an element of doubt and you foresee yourself as an inheritor of work not clearly assigned to a specific subcontractor.

Fan Coil Units

Factory-furnished controls are prevalent in commercial work. However, factory-*furnished* does not mean factory-*wired*. Electric fan coil units and their controls are frequently shipped to a job together, but all too often the controls arrive as loose components shipped with the fan coil units. Return air thermostats and fan speed switches are two good examples. Again, the equipment spec may say that fan coil units "must be provided with return air thermostats and unit fan speed switches." By shipping them loose with the units, the equipment supplier has complied with the spec. The low bidder on the control job may have been the one who didn't ask that all-important question: Are these fan coil unit controls going to be factory-mounted and -wired? You can easily find out by picking up the phone and calling the manufacturer's rep. If you discover that the control devices are to be shipped loose, ask your customer, the mechanical contractor, if you should include the controls' work in your quote. Then the mechanical contractor will be alert to the problem and will query your competitors about what they are including, and this will help keep you competitive at bid time. The more information you can bring out into the open, the more equal will be the level on which you are bidding against your fellow controls contractors, and the less likely are the chances that you'll have underestimated if you're low bidder or that you'll lose the bid if you include items not included by others.

Water Chillers

Large pieces of mechanical equipment represent many thousands of dollars in equipment cost in the mechanical contractor's bid. One big piece of gear that is a major expenditure on central cooling jobs is the water chiller. Sometimes a very small item is shipped with the chiller. That item is the chilled water flow switch, and it must be wired. Located in the chilled water piping to sense water flow, the flow switch tells the chiller when it is safe to operate. It does this by closing an electrical contact when flow is adequate. The flow switch is a small item, and you may want to ignore it in your quote if the control wiring part of your work is quite large anyway. But if the flow switch is not factory-mounted at the inlet-outlet piping of the chiller, you can be sure that it won't be factory-wired. Here again, check with the manufacturer's rep.

Exhaust Fans and Power Roof Ventilators

Exhaust fans and power roof ventilators can cost you money if they arrive on the job with electric actuators on their outlet or inlet louvers. Consider the case of a large warehouse with a dozen power roof ventilators featuring electrically actuated louvers. The fan motors themselves are 480-V three-phase motors, but a 120-V acutator is included in the power roof ventilator assembly. On the electrical floor plans, all that was shown was the three-phase power circuit that goes through the motor starters and on to the fan motors with local disconnects. No provision was made for the 120-V circuit needed to operate the louver to be interlocked with fan operation.

Look carefully at exhaust fan and power roof ventilator specifications and at fan schedules on the mechanical drawings to see if electric actuators are required. If you find yourself left with the job of wiring the actuators, instead of using long runs of 120-V circuit consider the alternative of furnishing and wiring 480- to 120-V transformers in each of the fan units so that the transformer is energized simultaneously with fan start-up, creating the 120 V necessary for operation of the 120-V louver motor.

Equipment Furnished by Others

To be an effective controls contractor, you must develop a lot of savvy about the equipment routinely found on commercial HVAC installations. Figure 12-1 shows what may appear to be two incompatible pieces of equipment, a heating-cooling thermostat and a cooling-only package rooftop unit, with its terminal strip that shows where the control wiring connects. Almost universally, terminal designations used in both low-voltage thermostats and mating low-voltage terminal strips in HVAC equipment share the same meaning. As shown in the figure, R indicates the positive, nongrounded leg of the 24-V transformer within the rooftop unit, and R within the thermostat denotes the termination point for the positive feed to the thermostat. Y represents the connection terminal for the control wire responsible for bringing on cooling, and G denotes the place for termination of the wire that brings on the indoor fan within the rooftop unit (or the evaporator fan in split systems). W indicates the termination point for the wire to institute heating, and C marks the terminal that connects to the other side of the 24-V control circuit transformer. In Fig. 12-1*b*, there is no W on the terminal strip within the cooling-only rooftop unit, and for good reason. There is no means of heat generation within the unit: It's a cooling-only unit. And yet these two components could very well have been

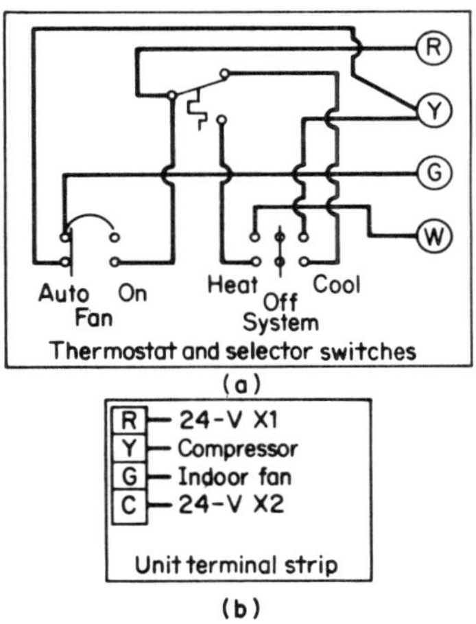

Figure 12-1 (*a*) Heating-cooling thermostat with switches and (*b*) cooling-only packaged rooftop unit.

furnished by the rooftop unit manufacturer in accordance with a specification that calls for a remote, duct-mounted hot water coil, intended to be zoned off of an existing hot water heating system using a small circulator to establish coil flow when heat is required. In compliance with an equipment specification, an equipment manufacturer might provide a rooftop unit, a heating-cooling thermostat, a hot water heating coil for duct mounting, and a circulator. And in the control section of the project specifications, the controls contractor is told to connect these factory-furnished components to provide a sequence of operation in which the rooftop unit's indoor fan and refrigerant compressor operate when there is a call for cooling and in which only the fan within the rooftop unit operates in conjunction with the hot water coil circulator when there is a call for heating. This is a classic example of the incidental supplemental control wiring required of the controls contractor in order to effect a sequence of operation using components furnished by others. Can you see some problems? As a controls contractor who is bidding for the work of connecting these pieces of equipment, you must visualize completely how the wiring will be accomplished in order to discover the additional controls and wiring needed to make it work.

When you look at this control scheme, the following items should come to mind. Recognition of the necessity of these items in the control scheme becomes vital when the scheme will be used in several places within a building.

Relays A 24-V relay will have to be provided, and its holding coil must be connected between W on the thermostat (Fig. 12-1*a*) and C on the terminal strip within the rooftop unit (Fig. 12-1*b*). Its normally open contact will have to be wired to energize the fractional-horsepower 120-V circulator when there is a call for heating.

The thermostat shown in Fig. 12-1*a* anticipates start-up of the indoor fan in the heating mode through the use of a plenum switch in a warm air furnace. Therefore, an additional contact on the 24-V heating relay will have to be used to close the 24-V circuit back to G to start the fan when heat is called for. Installing wire from thermostat termination point W through the normally open contact to thermostat termination point G will start the fan when the thermostat calls for heat. However, with the thermostat fan switch in the auto position, G will feed back to Y, causing the refrigeration compressor in the rooftop unit to run also. Therefore, a third, normally closed simultaneously energized contact must be used to open the connection between thermostat Y and rooftop unit Y whenever heating is called for in order to prevent the undesired operation of the refrigeration compressor. This third contact could be an additional 24-V relay with a normally closed contact, or it could be the third pole of a three-pole double-throw 24-V relay.

By identifying the characteristics of the control components furnished by others, you can prepare to complete the sequence control wiring requirements which will be needed when the system is put into operation. Therefore, it is important for you to be familiar with the HVAC equipment to be furnished by others so that you can prevent shortfalls in your estimate that will leave you holding the bag because of a spec that calls for a sequence of control without identifying all the necessary accessories. You must be able to integrate the sequence that you are required to develop with the specific control devices furnished by others. In a control job in which many control devices are furnished by others and in which the control wiring will be your responsibility, it is imperative to sketch out the wiring details and prove them on paper before making a commitment to perform the work. Sketching out the wiring or piping details of an electric control job or a pneumatic control job should be automatic anytime you are estimating. By regularly proving the scheme on paper before adding it to your estimate sheet, you will develop good estimating habits.

I believe that estimating should be centered on the development of schematics on multiple pages sequentially numbered and with appropriate subtitles to identify the various control loops. I always note on each schematic that it is typical for however many devices are involved. If the sketch is typical for five air-handling units, I write "typical for 5." I suggest doing this even to the point of writing "typical for 1" when only one is the case—just to reinforce the practice. This good estimating habit helps you keep track of the number of devices needed. As you transfer the devices and their quantities to estimate sheets, check off the symbols in your takeoff sketch representing the controls required with a red pencil. Then you'll have a record of your control system on paper, with check marks to assure that you took each device into account.

Cooling Towers A piece of equipment furnished by others that is easy to overlook is the cooling tower, particularly if the spec doesn't require the controls contractor to provide controls for it. And yet, the absence of the requirement should be your cue to look closely to see how capacity is to be regulated. There are various methods of controlling cooling tower capacity, and most are spelled out in the controls section of the HVAC specification. Typical methods are cycling of tower fans and modulation of water flow over the tower. However, cases do arise in which a cooling tower package is specified, and the cooling tower equipment spec calls for "factory-furnished tower capacity air flow dampers with thermostat." Now, this might be the only time that this item is mentioned in the bidding documents. It doesn't show in the control diagrams on the mechanical sheets, nor does the sequence for it appear in the control section of the specifications. On the electrical drawings, of course the three-phase power circuit to a weatherproof disconnect is shown, and the electrical contractor will estimate the power wiring. All seems right with the world. Six months later, however, when the cooling tower is sitting on the job, the customer's representative discovers (1) a proportional electric damper actuator that is linked, but not wired, to the tower capacity control dampers and (2) a proportional electric thermostat with separable socket immersion well that was shipped loose with the tower and that is meant to control the tower capacity dampers. The representative calls your customer's office. Your spec says, "controls contractor to provide necessary control and interlock wiring." Your customer calls you. Guess who your customer wants to wire the cooling tower condenser water thermostat and the damper actuator? What was the basis of your contractual agreement with your customer? Is it a purchase order which reads "per plans and specifications"? Did you recognize the need for this control wiring, and is it allowed for in your estimate?

Remote Condensing Units in Split Systems Last, consider the classic case of remote condensing units in split systems. In the equipment spec, the air-cooled compressor-condenser unit serving a large direct expansion coil in an inside air handler is described as operating in the pump-down mode when the call for cooling no longer exists. Two stages of cooling are planned for, and the job of the controls is to energize two liquid line solenoid valves located at the evaporator when there is a call for cooling. All the control wiring appears to be within the fan room, because the condensing unit will be shut down by its built-in low-pressure switch once both solenoids close. Built-in capacity controls enable the condensing unit to "see" the closed liquid line solenoids and allow the compressor responding to the appropriate solenoid to pump on until it pulls down line pressure and opens the low-pressure switch. Of course, the three-phase power circuit is shown on the electrical drawings, and the electrical contractor will run power to the outdoor unit. However, unknown to the controls contractor, a 120-V control circuit transformer was an option at extra cost, which the customer did not purchase. The outdoor unit requires 120 V for the factory-furnished controls to operate; it needs a control circuit. In addition, a hot gas bypass valve, electrically actuated, has been discovered inside the condensing unit in the compartment with the two semihermetic compressors. But the bypass valve is not wired. It needs to be wired in parallel to the first stage of cooling, the first liquid line solenoid valve. It looks as if the unit needs four wires extended to it from inside the building, two for 120-V power and two for energizing the hot gas bypass valve if the controls are calling for cooling. Invariably, control wiring of some kind is required between the indoor and outdoor units.

Over the years, when I've had the responsibility for control wiring and split systems were involved, I would make provision in my estimate for an empty 3/4-in raceway between the indoor and outdoor units, unless I could ascertain beyond a shadow of a doubt that no wiring between the two was required. This can be hard to do at bid time when you're not sure whose equipment will ultimately be used on the job. However, the conduit between the two pieces of equipment is much cheaper when it is planned on as a part of the original control wiring requirements than when it is installed after construction has advanced to a point where craftsmen in all trades are well into the project. Then if it develops that wires between the two are required, pulling them in and terminating them won't be a jarring economic shock to job cost—certainly it would be nothing like the shock that would occur if the conduit were not already there.

KNOWING YOUR JOB

Knowing your job from an electrical standpoint, then, consists of being aware, particularly as you prepare to bid on a job, of these and other areas of control projects. Examining the equipment specs when bidding the controls portion of a project will keep you aware of the possibilities of additional control wiring requirements that could surface after you are into the installation phases of the work. As a controls estimator, you must ask questions of yourself as you read the bidding documents, and you must make notes that you must follow up on with other suppliers before you bid the job. And when you quote on work in which you have excluded all control wiring other than that required for the devices that you are furnishing, it makes sense to see that the exclusion is also expressed in your customer's order to you. "Per plans and specifications" is not sufficient. Unless specifically stated otherwise, the responsibility for wiring controls in equipment furnished by others may well be yours even if you have not planned on it. On the other hand, if your customer wants you to field-connect components furnished by other manufacturers, you can estimate it and plan for it. Once it's in your estimate, you will have no reservations about assuming the responsibility for performing the work. But you must know what's required. Estimate it and plan for it, or exclude it, then make your position clear. It's just plain good business to protect your job costs.

13

Control System Electrical Subcontracting

For the controls contractor who prefers to do only the pneumatic portion of the temperature control system, subcontracting the electrical portion may present a convenient solution. In order for a profitable and lasting relationship to exist between the controls contractor and the designated electrical contractor, the work to be performed by the electrical contractor must be clearly presented and must be complete.

SUBCONTRACTING GUIDELINES

In soliciting electrical estimates for control work, follow these guidelines:

1. Lay out the job in plan-view form.
2. Specify responsibility for mounting devices and terminating wiring at control devices.
3. Be familiar with the electrical spec so that the type of raceway required can be shown on your drawing.
4. Cover *all* the work that your subcontractor will be required to do.
5. Present your request for pricing to your subcontractor early enough to allow adequate time for careful pricing.
6. Make it clear that you will reimburse your subcontractor for any work which develops during the course of the job and was not covered in your drawings.
7. Provide adequate supervision.
8. Present correct wiring termination diagrams.

Many electrical contractors steer away from control wiring because they lost money or failed to make a fair profit on control work. In such cases, usually the controls contractor did not see to it that the electrical subcontractor was well informed and sufficiently supervised. On the other hand, many electrical contractors enjoy good repeat business arrangements with controls contractors because of the confidence that the electrical subcontractors have in the performance of the controls personnel.

ADVANTAGES OF USING AN ELECTRICAL SUBCONTRACTOR ON A REGULAR BASIS

Some of the benefits of having one subcontractor work with you on a recurring basis are listed below:

1. The electrical subcontractor becomes familiar with the work you do and thus becomes increasingly efficient at doing your type of work.
2. You can prepare pricing guidelines for use in your own bidding on the basis of past costs for the subcontracting work.
3. Your subcontractor is more willing to estimate large jobs for you within the time frame your require to meet quotation deadlines.
4. Other electrical contractors doing power and lighting wiring on projects will be less resentful of the presence of a second electrical contractor on the job when they know that contractor routinely does your electrical control work.
5. And, of course, you are assured that experienced electricians are wiring your controls, making your job of supervision easier.

As a control system estimator, you need to be adept at packaging the bid materials for your electrical subcontractors. This means that your request for pricing from the subcontractor should be thorough and clear. The work you submit to your subcontractor for pricing should be presented in the same manner each time. Just as anyone who works with you on a continuing basis will become familiar with the work to be done, so too will the electrical subcontractor become familiar and at ease with the materials to be priced if you always prepare them in the same way. Thus the efficiency of the electrical contractor as estimator improves through repetition, as does that of the installers.

The consistent method of pricing used by the subcontractor provides the basis for developing pricing yardsticks that will help you anticipate what the electrical subcontractor's prices will be. For instance, the electrical subcontractor might always use unit pricing for different types of controls, and so you can use unit pricing when you work up control system costs. An electrical subcontractor who is comfortable with your work will usually agree to the various costs you have worked up for the wiring of specific control devices. You should periodically check the unit prices with the electrical subcontractor to see if they need to be revised. Unit pricing can add speed and accuracy to your efforts at costing out electrical work pertinent to controls in anticipation of the electrical subcontractor's price for performing this work.

You should approach the job of estimating large temperature control systems with care, both in assessing the involvement of your crew and in determining the extent of electrical wiring to be done. In this case, unit pricing may not represent a sufficiently accurate measurement of what your subcontractor's electricians will have to do. When you and your electrical subcontractor maintain a satisfactory relationship, your electrical subcontractor stands ready to help with pricing when needed. Because you both routinely profit from the installation of control systems, the electrical subcontractor makes room in a busy bidding schedule to be of service. Help in pricing is particularly necessary when there is a fairly heavy schedule of work to be bid, by either one of you. And since you both work together regularly, you can be sure that the electrical subcontractor will provide competitive pricing—pricing which compares quite favorably with the pricing you might receive from a subcontractor whom you approach only once in a great while.

Your consistent use of the same electrical subcontractor, makes people on the jobsite begin to think of you both as partners in the construction arena. Thus when the electrical subcontractor arrives at a project to perform control wiring, this comes as no surprise to the project's jobsite electrical contractor, and the relations between the electricians on the job are likely to be much smoother. This is very important because it helps to prevent problems of jurisdiction. For example, consider the electrical work that might be necessary in connection with an electrical motor starter. The controls contractor usually is responsible for bringing the motor starter control circuit to air-handling-unit safety devices such as firestats, freezestats, or smoke detectors. However, if the starter features a hand-off-auto (HOA) selector switch on the starter enclosure cover, responsibility is not so clear-cut. In the absence of any control devices, the holding coil control circuit wiring through the HOA selector switch would certainly be performed by the project jobsite electrical contractor. Where harmony is not present between the two electrical subcontractors conflict can arise over the wiring of the HOA selector switch, with the jobsite electrical contractor refusing to do it because all the control wiring is to be done by others and with the controls electrical subcontractor refusing to do it because the work is within the motor starter enclosure. The amount of work involved is insignificant to both of them, because both are involved with the wiring of the control circuit of the motor starter. An electrical subcontractor who works consistently with you will probably be quite comfortable with the traditional position of completing the control wiring to the cover-mounted HOA selector switch while making other connections in the starter enclosure that are relevant to the safety devices on the air-handling unit, and so the dispute would never surface in the first place. Thus an unnecessary dispute over what might otherwise be a gray area of responsibility need never become a jobsite issue, because the likelihood of resentment from the jobsite electrician is diminished almost to extinction when your subcontractor is well known as your associate.

It is necessary to supervise electricians during the connection of control devices to ensure that the installation is being made in accordance with the job drawings, particularly if the electricians are unfamiliar with the control equipment being used. When you continue to use the services of the same electrical subcontractor, the electricians who work for the subcontractor become comfortable with the control devices, because they get used to working with them. When this happens, your need to supervise them is reduced, thereby saving you time and enabling you and the electrical subcontractor to handle more work in the same time frame. The economies here become obvious, because to a contractor, time is money. Savings can be realized by you and your subcontractor as the electricians become more proficient with your devices.

RESPONSIBILITIES OF THE CONTROLS CONTRACTOR TO THE ELECTRICAL SUBCONTRACTOR

The eight guidelines offered at the beginning of this chapter are effective ways to encourage an electrical subcontractor's confidence in your work and to lay the groundwork that will permit the subcontractor to earn a fair return on the electrical work. The plan view of the electrical work enables the subcontractor to satisfactorily itemize amounts of conduit and wire and to count the devices to be mounted and wired. Controls contractors are the best qualified to make this plan view, showing the intended routing of conduit and wire as they make their way to the controls. Preparing a good plan view is one of the most important considerations in obtaining fair control wiring prices.

In looking at the control wiring, you may decide to purchase a raceway system only, with wire installed and tagged for terminations by you. Frequently, controls people prefer to make their own terminations and, in some cases, to mount the equipment. If this should be the preference, clearly note it on the plan view and be sure to request wire tagging and 6-ft lengths of flexible conduit complete with connectors. Once the raceway and wire are installed, you can then mount the devices and do the terminations yourself. An electrical subcontractor who often works with you might do the mounting and terminating, because the subcontractor's personnel have come to know the installation requirements of the devices.

Whether or not you choose to do the wiring and make the terminations, be sure to check the electrical section of the project specifications to determine what kind of raceway is required. You must pass this information on to your subcontractor because in all likelihood the subcontractor is not involved in the job other than with your work and so will not have access to the project specs.

The electrical subcontractor looks to you to recognize the full scope of the work to be done. All too often, control wiring requirements are overlooked and arise later on in the job, after the agreed-upon control wiring price has been established. You need to read the specifications thoroughly and be aware of wiring which can be construed as control wiring and thereby become your responsibility. Examples include chiller interlock wiring, remote condenser interlocking, tie-in of smoke detectors to fire alarm panels, and filter media runout interlocks—all invariably involving equipment you did not furnish. If your control wiring responsibilities were not clearly set out in your quotation to your customer and in the subsequent purchase order from your customer, you are likely to inherit additional work which has to be done, and so is your electrical subcontractor. Good controls contractors will show all the work to be done in accordance with their obligations to their customers.

We all appreciate having enough time to perform our work, whatever it might be, and electrical subcontractors are no exception. They will do their finest work when there is ample time in which to do it. Presenting the quotation requirements to them early in the game shows the subcontractors that you respect their time and appreciate the fact that they have other work to do also; moreover, it also sets the stage for a more careful consideration of your needs. Having to rush creates uncertainty, as insufficient time prevents a thorough understanding of the project. As a form of protection, a subcontractor will add extra dollars to the estimate to ensure adequate coverage—a contractor who is in doubt many well "double 'til the doubt disappears." Thus the price to you could be higher than it should be. As a result, the bid may be lost, or if the bid has already been won, there could be difficulty in obtaining a price consistent with what you have allowed.

The first few jobs you do with an electrical subcontractor can well influence the subcontractor's thinking about doing future work with you. It is wise to assure the subcontractor that any discrepancies between the proposed work and the actual work will be reimbursed. Control wiring is not always as straightforward as power and lighting wiring, and until the subcontractor is comfortable with controls work, such an assurance will encourage the subcontractor's continued participation at fair prices. When controls people have done their jobs well, they should have no reluctance in giving assurance.

As the work progresses, make periodic jobsite checks to prevent misunderstanding of your drawings and to show that the electrical subcontractor can count on you for adequate supervision. This makes a big difference in the subcontractor's labor costs, because the installers won't have to waste time second-guessing your intentions. Once the installers become efficient at doing the work, the subcontractor's pricing to you will assume predictable properties.

The work the subcontractor does will be no more accurate than the controls contractor's drawings. As academic as this seems, carelessness has been evident in the control diagrams of more than one well-recognized controls contractor, and it points up the need to have all submittals checked for accuracy before they are presented. Errors are costly, and the subcontractor knows this.

14

Summarizing the Temperature Controls Contract Bid

ACCURATE APPRAISAL OF COSTS

The quotation for the installation of a control system takes into account all the costs associated with the job, allowances for overhead, and a reasonable profit. If you win the contract, you must earn a profit. In order for a contracting firm to continue to exist, it must realize a reasonable recovery from its construction activities. For this reason, an accurate appraisal of installation costs is a prerequisite to the development of a correct bid price. Far too many firms have failed financially simply because they did not have a good grasp of what their costs were. Truly, you live or die economically by your estimate. Of course, sometimes contracts are won by those who simply bid too low, quite often through omission of one or more cost elements. In open-market competitive bidding, you are continually pitted against those who may be failing to properly account for all cost-related segments of a job. Even so, continue to systematically use good estimating practices for reliable predictions of job costs. Then, if necessary, narrow your recovery margins to win the installation awards. This way, your plan for the job will be valid, and it might be that tight engineering, aimed at cutting costs through more effective application of control equipment, can increase your projected recovery while assuring the customer of an effective, cost-efficient control system. It's far better to know that you've got a close job and that you must work to improve your margin than it is to expect wider margins which never develop. The first rule of an estimator is to estimate correctly and accurately. This allows management to make bidding and marketing decisions with the assurance that the cost basis is sound.

With the labor, materials, and subcontract costs in hand, you can begin to estimate the total raw job costs, from which overhead and profit margin decisions can be made.

THE BID SUMMARY SHEET

Form 14-1 is a sample bid summary sheet which permits line-by-line entry of job-related cost segments.

Cost of Control System Devices

On the form, the total cost of the required control system devices is addressed first. This figure includes the cost of all control components, such as thermostats, controllers, relays, transformers, time clocks, and control valves. Devices that are *field-mounted* in control panels should be included here as well, since Form 14-1 also provides for any special panels which must be ordered from outside vendors. Occasionally, it may be to your advantage to order panels, prepared from your drawings, that are completely prefabricated with terminal strips for field wiring. This may be especially true if your workload in the field is fairly heavy, if jobsite conditions are not favorable for on-site panel work, or if your facilities for in-house assembly are limited. The cost of control valves is included in the costs of equipment room devices because good lines of control valves with actuators are readily available from the major control manufacturers.

The cost of control dampers is listed on line 4. Listing this item separately facilitates the breakout of damper costs if customers prefer to purchase the dampers themselves or to manufacture their own dampers. Take care if you are doing the purchasing, because oftentimes damper specs mention specific brands and styles of dampers which tie you to published closeoff characteristics or linear flow patterns peculiar to certain manufacturers' dampers. Having a separate cost entry for dampers makes comparison of damper prices for the job much easier. Compressor and dryer costs are listed on line 5.

Materials Costs

The cost of rough-in materials, including tubing, wire, raceway, fittings, hangers, and accessories, which was estimated on Form 9-1 (the space rough-in estimate sheet), is entered next. Then the cost of the equipment room rough-in materials, from Form 3-1 (the equipment room labor and materials estimate sheet) is entered and appears as an allowance of assumed sizes for estimating purposes. Then the space rough-in materials cost and the equipment room materials costs are added together for costing. If the bid is won, then a definitive itemization of equipment room materials can be prepared.

As mentioned earlier, special panels may be needed for the job. If so, their cost can be listed on line 8 of Form 14-1. Panels constructed in a panel fabricator's shop require field connections only. If you are to provide the control components, enter their costs on the panel estimate (Form 10-1). This makes it much simpler to order panel parts later on, and when estimating and engineering are separate production activities, it helps the application engineer in understanding the intentions of the estimator. Now total lines 6 to 8 to arrive at the "Total materials cost."

Bid Summary Sheet

Job name ______________________ Date __________

Estimator ______________ Bids due ______________ Page ___ of ___

Devices and materials

Control system devices

1. Space devices costs	$ ______		
2. Equip. room devices costs	$ ______		
3. Panel devices costs	$ ______		
4. Damper costs	$ ______		
5. Compressor and dryer costs	$ ______		
Total devices costs	$ ______	$ ______	

Installation materials

6. Space rough-in materials costs	$ ______		
7. Equip. room materials costs	$ ______		
8. Special panel costs	$ ______		
Total materials costs	$ ______	$ ______	
Escalation allowance		$ ______	
Sales tax allowance		$ ______	
Total devices and materials costs		$ ______	$ ______

Labor

Labor

9. Space rough-in hours	______			
10. Space-mounting hours	______			
11. Equip. room hours	______			
12. Panel fabrication hours	______			
13. Travel time hours	______			
* 14. Checkout and supervision hours	______			
Total labor-hours	______			
____ hours @ $ ____ per hour		$ ______		
____ hours @ $ ____ per hour		$ ______		
____ hours @ $ ____ per hour		$ ______		
Total installation costs		$ ______	$ ______	
Labor travel expenses			$ ______	
Engineering costs			$ ______	
Taxes, insurance and fringe costs on labor			$ ______	
Total labor costs			$ ______	$ ______

Related costs

Peripheral costs

Bonds, cleanup, construction shack	$ ______	
Permit fees	$ ______	
Factory-mounting charges	$ ______	
Subcontracts	$ ______	
Submittal data	$ ______	
Warranty	$ ______	
† Technical specialist's allowance	$ ______	
Owner orientation allowance	$ ______	
Total peripheral costs	$ ______	$ ______

Total job raw cost		$ ______
______ % overhead	$ ______	$ ______
Job cost and overhead		$ ______
______ % profit	$ ______	$ ______
Job cost, overhead, and profit (bid price)		$ ______

*Special allowance only, if applicable
†Outside purchased assistance only, if applicable

Form 14-1 Control systems bid summary sheet.

Special escalation rates should be applied to projects that will take longer than a year so that increasing costs won't affect the accuracy of the estimate. Probably 6 percent a year is the minimum which should be used. Take the foregoing "Total devices costs" and "Total materials costs" and add them together and enter 6 percent of their total on the escalation allowance line. The escalation allowance is added to the total cost and becomes part of the "Total devices and materials costs." Sales tax on devices and materials should be entered as applicable according to the tax laws in your area. The "Total devices and materials costs" should cover everything the installers need in order to proceed with the work involved in the project. All costs below this point on Form 14-1 will relate to labor costs, subcontract costs, and costs that are incidental to making the job an active project.

Labor Costs

While the costs of devices and equipment can generally be predicted with good accuracy, the cost of labor does not usually come as close in actuality to what was estimated. For this reason, different variables are provided for on the labor estimate sheets to help nail down the differing conditions which cause labor production to vary. Once you are ready to enter the labor-hours on Form 14-1, you have already thoroughly explored job conditions, techniques, and any special difficulties and you have already made decisions.

Since the entries in this next section in Form 14-1, "Labor, " have proved over the years to be the ones most likely to deviate from the anticipated, the installation plans from which the labor-hours evolved need to be well documented and conscientiously passed on to the field supervisors. The entries should be realistic enough to permit the work to be fully accomplished by reasonably productive personnel and, at the same time, should represent the best laborsaving techniques that you could foresee for the project. The entries should be conservative and yet flexible enough to present a laborsaving challenge to your installers. Too tight an entry will not permit the field people to complete their work in the time required, and thus will reduce their enthusiasm to beat the estimate. But a well-planned estimate that allows a sufficient number of hours for the work enables the technicians to make a showing and thus receive the well-deserved commendations due efficient persons. Therefore, when you enter your labor-hours, satisfy yourself that they possess the right allowances for those who work for you and want to turn in their best performance and have it recognized.

Travel Time Travel is an allowance in hours for road time to and from the job. To find the number of hours, divide the job labor-hours by workers' total hours per day to determine the number of round-trips for the job. Multiply the number of trips by the time required to travel both ways. Take your answer and multiply it by the number of workers who will be traveling. This gives you total travel hours.

When the job is farther away than 75 mi, consider having your crew work four 10-h days with three nights' worth of out-of-town expenses, which reduces travel time to one round-trip a week. In this case, enter the labor-hours for weekly round trips on line 13. Figure travel allowances both ways, and choose the most economical. Labor, travel, and expenses go hand-in-hand, so installation time allowances must be as realistic as possible.

Checkout and Supervision On line 14, the "Checkout and supervision hours" line, enter the time that you estimate will be required in addition to installation hours, which already include commissioning the control devices. Labor-hours for checkout and supervision are sometimes needed for people on your work force who have special knowledge about highly technical equipment, such as certain solid-state energy management equipment or electronic interface equipment used for building-to-building data transmission. On this line you can also list the hours that you may have to spend on the job merely because you are more familiar than anyone else with a particular device—a device with which others on your work force may not feel as comfortable. Consider the commissioning of the device a training session and enter as much time as you might need for providing your workers with the proper training.

Total Labor-Hours The rates per hour of labor vary from area to area and also vary over time. For this reason, there are three lines in Form 14-1 on which you can list varying quantities of hours at differing rates; this gives you an on-sheet accounting of how you expect time to be apportioned within any given rate period.

Travel Expenses Labor travel expenses are the personal expenses for which workers must be reimbursed when they are out of town on jobs. In estimating this item, consider the number of people involved and the number of days they must be away and enter the appropriate allowance on the "Labor travel expenses" line to cover their job-related personal costs.

Engineering Costs Engineering costs cover those administrative and technical functions that are required for delivering the submittal and ordering the materials. From the plans and specifications, detailed diagrams of the job must be drawn up for approval by the consulting engineer and for use by the installers. The labor for drawing up the diagrams for a job should be charged into that job and should be covered by an estimated allowance of engineering time. Engineering expertise and efficiency vary from shop to shop, from the large office, where many people do the work, to the one person shop, where one individual lays out the work and installs it as well or subcontracts it. But adequate time must be planned for, because in contracting, time is money. Normal allowances for engineering time frequently are expressed as a percentage of total labor *installation* costs. This excludes travel expenses, as it should, and focuses on true labor installation costs only. I suggest an entry of 10 percent of total labor installation costs. Thus, if total labor installation costs were $6500.00, the entry on the engineering line would be $650.00. On the basis of your experience, feel free to enter whatever percentage of labor installation costs that you feel is most accurate, but 10 percent is listed as a standard, suggested percentage. In many instances, such an allowance has been shown to closely approximate the engineering time actually required for making the estimate into a working project drawing, complete with approvals.

Taxes, Insurance, and Fringe Costs Finally, an allowance must be made for payroll taxes, insurance, and other fringe costs. Frequently this allowance is entitled "burden." "Burden" is a term that includes those costs that are related to the maintenance of a labor force, and it is expressed as a percentage of the total job labor-hour costs. The percentage you use will depend on your particular situation, but 16 percent is usually considered a reliable rate to cover the costs of fielding a work force. Enter 16 percent of the total installation labor costs (or a percentage of it that you know to be suitable for your area) on the "Taxes, insurance, and fringe costs on labor" line.

Peripheral Costs

Toward the bottom of Form 14-1, peripheral costs are entered.

Bonds, Cleanup, and Construction Shack Performance bonds, when required, are direct job costs that can be entered here.

On a large project, peripheral costs might include hiring workers to clean up job debris. You would not want to use the more expensive time of installers or technicians for cleaning up, and so the cleanup allowance permits you to hire outsiders, usually, to do this work.

Also on a large job, you might require an area for storing equipment and keeping job drawings on the jobsite. In this case, it is not unusual to ask the general contractor to build you a construction shack, for which you will issue a purchase order. Or if you prefer, you might rent a trailer for jobsite use. Either of these alternatives would be entered as a peripheral cost.

Permits Certainly, permits are required for electrical work, and in some areas permits are required for pneumatic piping work too. On the "Permit fees" line, you would account for such anticipated costs.

Factory Mounting Terminal unit manufacturers often prefer to mount and connect terminal unit controls in their factories. In this way they can be sure that the mounting and connecting are done in the manner they prescribe. Usually, mounting the devices in the factory is more efficient than mounting them in the field, especially when the units do not have large control compartments. The control contractor provides the controls to the factory for mounting. Unit manufacturers will usually inform controls contractors what their mounting charges are, or the manufacturers might quote the charges to the mechanical contractor who will be buying the equipment from them. Sometimes the mechanical contractor, who has received the quote from the terminal unit manufacturer, expects the controls contractor to pick up this cost. When you are expected to pay for factory mounting, use the "Factory-mounting charges" line to list the charges. If the mounting charges are included in the cost of the equipment to the mechanical contractor, you don't need to make an allowance for factory-mounting charges, but check it out to make sure.

Subcontracts For electrical work which you elect to give out to a subcontractor, enter the dollar cost of the subcontractor's estimate on the "Subcontracts" line. In addition, include any other subcontracted work (pipeline taps, special painting, sheet metal, etc.) on this line also. Make sure the subcontractor's prices are complete and that you both understand what the subcontract will provide for. Also, plan on quickly awarding the contract if you win the job, and make sure that the subcontractor's guaranteed acceptance period permits enough time for all contractual awards to take place. As a controls contractor, you will occupy the third tier in the flow of downhill paperwork from the general contractor, and your subcontractor will be the fourth tier. If the quote from the subcontractor is firm for only 30 days, you may need to make sure the subcontractor is prepared to staff the job and handle the risk for the intended duration of the project. Sometimes a short acceptance period is indicative of only slight interest in doing the control wiring work for you. Be certain of your subcontractors, their contracts, and their abilities to perform for you. For those subcontractors with whom you work repeatedly, this of course need not be a concern.

Submittal Data Once you have received an order for a job, you will have to produce a submittal for examination by the consultants. The submittal consists, in part, of data sheets on the devices you propose to furnish. The cost of the data sheets should be considered a job cost because usually they must be purchased when large quantities are necessary. It is a good idea to keep some data sheets on hand at all times, since submittals can then be prepared without having to wait for a delivery of data sheets. Not all jobs require submittal literature, but for those that do, an entry should be made on the "Submittal data" line to ensure that the costs relative to obtaining control device literature are fed back into the project for which they are intended. If you prefer not to stock data sheets, you should order them as you need them; there's no reason why, once you receive an order and before the job is engineered, you can't skim your estimate sheets for model numbers so that you can order the necessary data sheets. The purchase orders for the data sheets would then bear the appropriate account numbers, thus assuring that the costs will be charged into the project. It is true that you usually receive a data sheet with each device you order, but you need the sheets before the orders are issued. You would do well to have your field people hold and return data sheets and perhaps you could save them for future use in submittals. In any case, getting the charges into the job should rightly be a concern for you.

Warranty The warranty period is that period, usually 1 year, during which you agree to take care of failures in workmanship or materials at no additional cost to the owner. You need to make an entry of perhaps 5 percent of raw materials and labor costs to cover trips back to the job for adjustments, replacements, or corrections in work or devices for that normal 1-year period after the job is turned over to the owner. It stands to reason that you will be involved at one time or another during the warranty period. Problems do arise, and parts do need to be replaced from time to time.

Occasionally, the job specifications may call for a warranty period longer than the ordinary 1 year, and if this is the case, an increase in the warranty dollar amount entry is in order. Consider adding 10 percent instead of 5 percent when a 2-year warranty is called for. You will have to be the judge of what percentage to include, depending upon the complexity of the system and the sophistication of the controls. However, in the great majority of cases, the 1-year term will prevail, and the 5 percent inclusion should be satisfactory.

To appreciate the validity of the warranty provision on the estimate sheet, consider that during the preparation of labor plans for your regular in-house work, you no doubt will allot time for your workers to handle warranty period callbacks. And in most cases, these calls will be handled by your best-qualified workers, who probably are the highest paid as well. By estimating some time for this work during the course of bidding the job, you will help prevent a shortfall in calculated labor-hours in your job as a whole, which would show up in your job operating summary as excessive labor costs.

Technical Specialists Sometimes you will need a technical specialist, and the next item on Form 14-1 allows for this. Will some of the devices demand special expertise in commissioning the instrument? Will you be required to call for outside technical assistance for specially purchased controllers? Do your special instrumentation suppliers include jobsite technical setup? The "Technical specialist's allowance" line serves as a reminder for you to reflect on the nature of the devices slated for use that may fall outside the scope of routinely furnished equipment. This line provides a place to include dollars which may have to be spent for outside technical help, for example, when there are controls in the estimate which are unfamiliar to your setup people. In this case, it may be helpful to note, perhaps in the margin, where you anticipate purchasing the outside technical assistance. Allocating the cost of technical assistance is appropriate if you can foresee the need for it.

Owner Orientation Once the control system is installed and the owner's personnel are prepared to take over the building, who will explain the control system and make sure they know how everything operates? You might do it, but probably your time as an estimator would be better spent in the administrative functions, which, of necessity, you look after to keep production running smoothly. More likely than not, someone else familiar with the work, perhaps your job supervisor or lead installer, will conduct owner orientation. Owner instruction is a part of almost every job, and so plan on allocating labor dollars for it. Sometimes the specifications call for a certain period of time to be spent on the owner's behalf, and when this is the case, the corresponding labor dollars can be directly entered onto the "Owner orientation allowance" line. Even if a designated time frame is not called for, most certainly instruction of the owner's representatives will be spelled out in some manner. The owner orientation labor dollar allowance prepares you for job costs in this area.

Total Peripheral Costs These peripheral, or support, costs in this third portion of Form 14-1 can now be totaled and prepared for inclusion with raw materials costs and raw labor costs, which will produce a total raw job cost figure. Once you arrive at that number, you have a prediction of what you can expect your installation of the project to cost. With your best efforts, you have documented the costs which can be anticipated should you win the bid.

The final steps now concern application of overhead and profit percentages to create that margin you must have to (1) pay your overhead and (2) recover a profit for the business. Your overhead markup is the first of two markups which represent your total job gross margin. Only you are in a position to know what your overhead markup must be. The size of your business, your annual work volume, the quality of work in-house, your fixed expenses, debt retirement, and other considerations which you face daily in the operation of your contracting business must come into play as you begin the first stage of job cost markup. You must allow for sufficient and realistic overhead. With the application of a well-thought-out overhead markup rate, you will arrive at the lowest price you can expect to quote without losing money on the job. You wouldn't produce profit at this point, but you would be able to pay expenses, keep the lights on, and keep your crew working. You already have a feel for your overhead undoubtedly, and so this rate is up to you.

Next you calculate a markup that will permit you to make a profit from the job; a profit is necessary to create work incentives for your top administrators and to allow for expansion of your business. In this matter, the bidding pattern in your area will give you insight into how much of a profit you can reasonably expect. As a business person, you are entitled to some profit. It is a necessary ingredient of an ongoing business. Profit is a percentage of overhead and raw job costs added together. A job with a raw cost of $10,000 and an overhead rate of 25 percent would produce a subtotal, after overhead is added, of $12,500. A profit of 10 percent, then, would add another $1250, for a bid price of $13,750.

Peripheral costs do not constitute major costs, but they must be taken into account. They are included on Form 14-1 more as reminders to consider each item in case an entry for it would be appropriate, thereby ensuring that the cost has not been left out. You will know which items apply and which ones don't. In discussing these costs as I have, I did not want to create an area of cost concern for you that doesn't exist. By the same token, I did not want to be negligent in failing to mention the significance of each. It is important that you do not include any cost that is not likely to occur. You must remain as competitive as possible. The objective of Form 14-1 is to address the probable areas of cost so that you, as the job estimator, will see that allowances are made as required, but only to the extent that is necessary. Your estimate should be trim, not fat, in order for you to be competitive. But it must include all the costs so that you have the best possible data upon which to make your marketing decisions at bid time. You need to know your costs and then proceed as you must.

15 • Temperature Control System Estimating Exercises

ESTIMATING AN ELECTRIC CONTROL SYSTEM

By now, you have examined the four estimate sheets presented in this book:

- The equipment room labor and materials estimate sheet (Form 3-1)
- The space rough-in estimate sheet (Form 9-1)
- The control panel estimate sheet (Form 10-1)
- The control systems bid summary sheet (Form 14-1)

It is now time to use them to estimate a sample job, so that you can become familiar with their application. The sample job is the installation of controls for water temperature in air-to-water heat pumps (Fig. 8-1), discussed previously in Chap. 8. In addition to installing the controls for the water temperature, the thermostats furnished by the heat pump manufacturer must also be mounted and wired, and a relay must be installed in each unit for shutdown via the safety-dependent circuit for protection of the heat pump when water flow is lost or excessive temperatures occur in the water loop.

In this sample job the control panel estimate sheet wouldn't be used. Since work out in the spaces of a building is usually done before equipment room work, you should first consider the areas of the building that will be occupied. Then you can plan the work to be done in the equipment room, where later you can mount and wire devices without fear that the space you are working in will be covered up before you can finish.

Implementing the Space Rough-In Estimate

In the sample job, all the control devices except the heat pump thermostats and the safety-dependent relays are located in the mechanical equipment room.

Space Rough-In Estimate Sheet

Job name XYZ Building Date ______

Estimator Harry Edwards Page 1 of 3

Area or floor	Feet	Size	Material	Cost/ft, $	Total cost, $	Labor/ 100 ft, h	Total labor, h
1st floor stats	300	1/2	EMT 4 Wires	0.55	165.00	6.2	18.6
2nd floor stats	300	1/2	EMT 4 Wires	0.55	165.00	6.2	18.6
1st floor relays	500	1/2	EMT 2 Wires	0.39	195.00	6.0	30.0
2nd floor relays	500	1/2	EMT 2 Wires	0.39	195.00	6.0	30.0
Total					720.00		97.2

Transfer space rough-in material costs to bid summary sheet, line 6
Transfer space rough-in hours (X 1.3 for high work) to bid summary sheet, line 9

Space devices, stats, sensors, etc.

Area or floor	Qty.	Model	Cost/ea., $	Total cost, $	Labor/ea., h	Total labor, h
1st floor	10	Stats	NA	NA	1	10
2nd floor	10	Stats	NA	NA	1	10
1st floor	10	Relays	10	100	1	10
2nd floor	10	Relays	10	100	1	10
Total	40			200		40

Transfer space device controls costs to bid summary sheet, line 1

Transfer space mounting hours X 1.2 difficulty factor,* if appropriate, to bid summary sheet, line 10 *Ladder required

Space device, new work: connect, mount, and calibrate, 1 h each
Space device, old work: connect, mount, and calibrate, 1.2 h each

Figure 15-1 Space rough-in estimating exercise no. 1.

Since the evaporative cooler is usually positioned just outside the mechanical equipment room, it is listed as part of the equipment room work. On the space rough-in estimate sheet (Form 9-1), the various controls and their costs, labor-hours, and tube or wire (in this case, wire) and EMT conduit are entered, as shown in Fig. 15-1. You can also list equipment furnished by others that you must mount and/or wire, but in this case you would not show equipment cost.

In a specification which gives all the responsibility for control and interlock wiring to the controls contractor, the thermostats furnished by a heat pump manufacturer will be wired by you. The thermostats will be entered on the space rough-in estimate sheet, and this is where you begin your takeoff. Assume that there are 20 heat pump units of the air-to-water type and that there are 2 floors in the building (let's call it the XYZ Building), with 10 units on each floor. Enter the quantities of thermostats (20), by floor and model, on the space rough-in sheet. Allowing an average run of 30 ft from each thermostat to each respective heat pump unit, you'll find a requirement for 300 ft of 1/2-in EMT per floor for thermostats. The manufacturer's installation information shows that four wires are required from each thermostat to each heat pump unit, and so you must enter the price per foot for EMT, fittings, boxes, and hangers, with wire included. At $40.00 per 500-ft roll of no. 14 copper THHN, the per-foot cost of wire equals 8 cents. Since you need four wires, you have to multiply the 8 cents per foot for one wire by 4, which gives you 32 cents—the cost per foot for four wires. EMT, at $15.00 per 100 ft, costs 15 cents per foot for the conduit alone, but fittings, boxes, and hangers are required and so the per-foot cost of 15 cents for conduit is multiplied by 1.5. Now the cost per foot is 23 cents and includes EMT, fittings, boxes, and hangers. The cost per foot of 32 cents for four wires and the cost per foot of 23 cents for EMT and fittings totals to 55 cents per foot for EMT and fittings with four wires. You are now ready to enter this unitized per-foot cost of 55 cents for raceway and wires on the space rough-in estimate sheet and then multiply it by the footages required as shown in Fig. 15-1.

For the thermostats, 6.2 h per 100 ft is allowed for the installation of 1/2-in EMT with four wires, and so 6.2 is entered in the "Labor/100 ft" column. Since 300 ft of 1/2-in EMT with four wires is required per floor for thermostats, the total number of labor-hours for thermostats for the first floor is 18.6 h and for thermostats for the second floor it is also 18.6 h.

The next items to consider when entering the labor and materials requirements for space rough-in are the safety-dependent relays that you must furnish and mount in the individual heat pumps. All of them are controlled identically and simultaneously by the safety circuit, which is connected in series through low-limit controller LL-1, flow switch FS-1, and high-limit controller HL-1. Thus a two-wire circuit extended to each heat pump is required. By scaling the conduit required, you discover that each floor needs 500 ft of EMT and two wires. This figure should include enough EMT and wire to reach the unit's compartment, where you will mount the relay. All EMT to this point is concealed.

Applying the same logic you used to figure your per-foot price for EMT and four wires, you find that 1/2-in EMT with two wires carries a per-foot cost of 39 cents (8 cents per foot for wire times 2 equals 16 cents per foot for two wires, plus 23 cents per foot for 1/2-in EMT with fittings, boxes, and hangers). At 500 ft for each floor, EMT and wire costs are $195.00 per floor.

A requirement of 6 labor units, or 6.0 labor-hours per 100 ft, is allocated for laboring the conduit with two wires. With rough-in now figured for both floors in the occupied spaces, you add the "Total cost" column and the "Total labor" column and find that $720.00 worth of rough-in materials is necessary and that 97.2 labor-hours will be needed to install them. Prices per foot for EMT and wire must be your decision which you should base on the prevailing costs at the time of the estimate. You can unitize any combination of conduit and wire on the basis of local costs at the time and make a reference chart for your own use in applying the per-foot costs to this method of estimating.

The lower part of the space rough-in sheet is for costs and labor-hours for the devices you are to install in the occupied spaces. Because the thermostats (abbreviated "stats" in Fig. 15-1) are being furnished by the heat pump equipment manufacturer, you should enter "NA" (not applicable) or a dash in the "cost/ea." column, to indicate that these devices are being furnished by others and that you don't have to pay for them. But you do have to install them, and your labor entry must allow for connecting the thermostats, mounting them, and terminating the four wires in the heat pump on the appropriate terminals. In Fig. 15-1, 1 h is allowed for each thermostat, and this includes making terminations at both ends of the wires coming to the thermostat.

The other devices located in space are the relays, which will be mounted within the heat pump control boxes and then connected to the wires pulled in from the safety-dependent circuit loop. The coils of the relays will be connected to the two single wires in the safety-dependent circuit, and the normally open contacts of the relays will break their respective heat pump control circuits within the unit to effect shutdown, if necessary. For mounting and connecting each relay, 1 h is allowed. Remember, access panels must be removed and reinstalled.

Totaling the "Total cost" and "Total labor" entries in the lower section of the space rough-in estimate sheet results in material costs of $200.00 for relays and a time of 40 labor-hours for installation of thermostats and relays. Notice that the costs of space devices and space rough-in materials are to be transferred to lines 1 and 6, respectively, of the bid summary sheet for this sample job (Fig. 15-3), which appears later in this chapter. The labor-hours for space rough-in and space mounting are to be transferred to lines 9 and 10, respectively, of the bid summary sheet.

Implementing the Equipment Room Labor and Materials Estimate

Referring to the control systems sketch shown in Fig. 8-1 during takeoff and transfer of data to the equipment room labor and materials estimate sheet for this sample job (Fig. 15-2) will help ensure that you account for all devices. In Fig. 15-2, the devices are listed as shown in Fig. 8-1. In actual practice, it's best to list the complete model number of the specific control which you are estimating. Later, materials can be ordered directly from your estimate sheets as long as accurate and complete model numbers are entered.

Equipment Room Labor and Materials Estimate Sheet

Job name XYZ Building   Date ______

Estimator Harry Edwards   Page 2 of 3

Room or unit number Mech. equipt. room   Floor or area ______

Control	Pneu.	Elect.	Qty.	Cost/ea., $	Total cost, $	Labor/ea., h	Total labor, h	A	B	C	D	E
T-1		x	1	45.00	45.00	2.5	2.5	1				
TC-1		x	1	285.00	285.00	3.5	3.5	1				
DM-1		x	1	—	—	4.0	4.0	1				
FMS		x	1	—	—	3.0	3.0	1				
PMS		x	1	—	—	3.0	3.0	1				
T-2		x	1	50.00	50.00	2.5	2.5	1				
TC-2		x	1	285.00	285.00	3.5	3.5	1				
V-1		x	1	350.00	350.00	5.0	5.0	1				
LL-1		x	1	60.00	60.00	2.5	2.5	1				
FS-1		x	1	40.00	40.00	2.5	2.5	1				
HL-1		x	1	45.00	45.00	2.5	2.5	1				
Panel rings	—	—				One		—	—	—	—	—
Total			11		1160.00		34.5	11				

Transfer total equip. room control device costs $ 1160.00 to bid summary sheet, line 2

Transfer total equip. room hours 34.5 (X 1.3)* = 34.5 to bid summary sheet line 11

(A) Elect. controls wired in EMT: 11 X 30 ft/ea. = 330 ft

330 ft EMT and wire (X 1.3)* = 330 ft X $ 39.00 /100 ft = EMT and wire $ 129.00

(B) Elect. controls wired rigid: ____ X 30 ft/ea. = ____ ft

____ ft rigid and wire (X 1.3)* = ____ ft X $ ____ /100 ft = rigid and wire $ ____

(C) Pneu. controls piped in EMT: ____ X 30 ft/ea. = ____ ft

____ ft EMT & tubes (X 1.3)* = ____ ft X $ ____ /100 ft = EMT and tubes $ ____

(D) Pneu. controls tube bundled: ____ X 30 ft/ea. = ____ ft

____ ft tube bundled (X 1.3)* = ____ ft X $ ____ /100 ft = tube bundled $ ____

(E) Pneu. controls tube copper: ____ X 30 ft/ea. = ____ ft

____ ft tube copper (X 1.3)* = ____ ft X $ ____ /100 ft = tube copper $ ____

Transfer total installation material costs to bid summary sheet line 7 = $ 129.00

* Use factor only if area over 750 ft^2

Figure 15-2 Equipment room labor and materials estimating exercise no. 1.

Control valve V-1 is an engineered item and must be sized on the basis of the gpm flow and system head pressures expected. In the controls diagram (Fig. 8-1), the valve is installed in a mixing application, and so closeoff is not likely to be a difficult matter. Costs of controls are based on prices charged by your normal supplier, and the hours are taken from Table 3-1, which lists the equipment room labor-hours. T-1, T-2, LL-1, and HL-1 belong to the rigid-stem sensor or thermostat family, because they are pipeline-mounted devices that sense pipeline water temperatures. Because the wiring to be used is EMT, each of the four devices is assigned 2.5 labor-hours. TC-1 and TC-2 are responsive to their remote temperature sensors and belong to the receiver controller family. Since they will be wired with EMT, each is assigned labor-hours of 3.5 from Table 3-1.

Damper motor DM-1 at the evaporative cooler was shipped loose in this case by the evaporative cooler manufacturer. Therefore, no cost is listed in the "Cost/ea." column on the equipment room labor and materials estimate sheet, but 4.0 labor-hours is selected from Table 3-1, as is appropriate for an electric damper motor wired with EMT.

The control wiring will take you into fan motor starter FMS and pump motor starter PMS for the evaporative cooler. The pump for recirculating sump water over the tubing in the cooler is pump P-2. Even though you are not furnishing the pumps or motor starters, you must wire into their control circuits. Because the motor starter is a contractor-type device with electromechanically controlled contacts, it is part of the relay family. It is furnished, mounted, and power-wired by others. In Table 3-1, 3.5 h is allocated in the "EMT elect." column for wiring a relay, but since you don't have to mount it, drop 1/2 h from the time allowed. Paddle-type water flow switch FS-1 is assigned 2.5 h. All equipment room controls are now listed on the equipment room labor and materials estimate sheet, and you can total the columns to get the total cost of devices and the total labor-hours required for the equipment room. The total costs for equipment room controls are to be transferred to line 2 of the bid summary sheet for this sample (see Fig. 15-3), which appears later in this chapter, and the total labor-hours needed for the equipment room are to be transferred to line 11. Multiplying by a factor of 1.3 is unnecessary because the equipment room is less than 750 ft^2 in area. An in-room allowance of 30 ft per device is built into the equipment room labor and materials estimate sheet, and so on line (A) of the sheet, you enter 11 devices at 30 ft, arriving at 330 ft of 1/2-in EMT with two wires. Remember from your work on the space rough-in sheet that 1/2-in EMT at $15.00 per 100 ft multiplied by 1.5 to include fittings, boxes, and hangers equates to 23 cents per foot. Two wires of no. 14 copper THHN at $40.00 per 500 ft come out to 8 cents per foot, and so two wires make 16 cents per foot. Adding the cost of wire to the 23 cents per foot for EMT equals 39 cents for 1 ft, or $39.00 per 100 ft. This cost multiplied by 330 ft yields $129.00 for equipment room installation materials costs. This figure is to be transferred to line 7 of the bid summary sheet for this sample job (see Fig. 15-3). Now you are ready to use the bid summary sheet to arrive at the price you plan to quote for the installation of the controls in the XYZ Building. Try to use your experience on past jobs when making allowances for items on the bid summary sheet.

Implementing the Bid Summary Sheet

Note that the items on lines 1 through 12 of the bid summary sheet for this sample job (Fig. 15-3) have been transferred there from other estimate sheets.

Bid Summary Sheet

Job name XYZ Building Date ____

Estimator Harry Edwards Bids due ____ Page 3 of 3

Devices and materials

Control system devices

1. Space devices costs	$ 200.00		
2. Equip. room devices costs	$ 1160.00		
3. Panel devices costs	$ -		
4. Damper costs	$ -		
5. Compressor and dryer costs	$ -		
Total devices costs	$ 1360.00	$ 1360.00	

Installation materials

6. Space rough-in materials costs	$ 720.00		
7. Equip. room materials costs	$ 129.00		
8. Special panel costs	$ -		
Total materials costs	$ 849.00	$ 849.00	
Escalation allowance		$ 133.00	
Sales tax allowance		$ 94.00	
Total devices and materials costs		$ 2436.00	$ 2436.00

Labor

9. Space rough-in hours	97
10. Space-mounting hours	40
11. Equip. room hours	35
12. Panel fabrication hours	-
13. Travel time hours	22
*14. Checkout and supervision hours	-
Total labor-hours	194

97 hours @ $ 15. per hour	$ 1455.00		
97 hours @ $ 8. per hour	$ 776.00		
____ hours @ $ ____ per hour	$		
Total installation costs	$ 2231.00	$ 2231.00	
Labor travel expenses		$	
Engineering costs		$ 170.00	
Taxes, insurance and fringe costs on labor		$ 384.00	
Total labor costs		$ 2785.00	$ 2785.00

Related costs

Peripheral costs

Bonds, cleanup, construction shack	$ -	
Permit fees	$ 50.00	
Factory-mounting charges	$ -	
Subcontracts	$ -	
Submittal data	$ 25.00	
Warranty	$ 261.00	
† Technical specialist's allowance	$ -	
Owner orientation allowance	$ 300.00	
Total peripheral costs	$ 636.00	$ 636.00

Total job raw cost		$ 5857.00
15 % overhead	$ 879.00	$ 879.00
Job cost and overhead		$ 6736.00
10 % profit	$ 674.00	$ 674.00
Job cost, overhead, and profit (bid price)		$ 7410.00

*Special allowance only, if applicable
†Outside purchased assistance only, if applicable

Figure 15-3 Control systems bid summary estimating exercise no. 1.

On line 13, look at the labor-hours that you will have to pay your crew to travel to the job. For purposes of this estimate, assume that the job is an hour's travel from the office and that you will pay your workers' travel one way. For this job, you plan to work a crew of two, a mechanic and a helper. The two must travel to the jobsite every day that work is required there of them. Totaling the hours on lines 9, 10, and 11, you find that 172 h of work is estimated. Each workday represents 16 labor-hours with a crew of two. And so 172 labor-hours divided by 16 labor-hours per day gives you 11 trips to the job, and allowing each person travel one way means that the two workers will use 22 h in travel for this job. With the 22 h entered on line 13 and added in, you now have a total of 194 labor-hours. Assuming that rates will not change during the job, you can allow half of the hours at the mechanic's rate and the other half at the helper's rate—rates of $15.00 per hour for the mechanic and $8.00 per hour for the helper have been assumed here. Multiplying the number of hours by the hourly rates and then adding your answers gives you $2231.00 worth of total installation costs.

But to to get a true picture of labor costs, engineering and burden must be considered. Engineering includes preparing the shop drawings for field installation and for submittal approval and ordering materials. Using the anticipated installation labor-hours as a yardstick to measure the scope of the job, you can apply a factor of 10 percent to give you the number of hours that you expect engineering to require. Looking at lines 9, 10, and 11 only, you can see that 172 h of active installation work is forecast. Multiplying 172 by 10 percent gives you 17 engineering hours, and at a cost of $10.00 per hour, $170.00 is allowed for engineering costs.

"Burden" is a term applied to the incidental costs of having people on the payroll, and it covers payroll taxes, insurance, workers' compensation, unemployment tax, fringe benefits, etc.. In actual practice, your situation will dictate the percentage to be used, but for purposes of this sample job, 16 percent is used. Installation and engineering costs are added together to obtain $2401.00 at the total job-assignable labor cost, and 16 percent of this is $384.00.

As discussed in Chap. 14, peripheral costs are the costs of items such as subcontracts, cleanup, bid bonds, and a construction shack. They are directly affected by the size of the job, and whether they apply depends on the job spec, subcontract requirements, and local regulations. When such costs are incurred on a job, they are rightly considered direct job costs and must be entered on the bid summary sheet. In the case of the installation of the air-to-water heat pumps shown in Fig. 8-1, certainly no entry is required for cleanup or for a construction shack. Electrical work usually requires a permit, and often the cost of the permit is based on the cost of the electric wiring. Fees vary from locality to locality, but for purposes of this sample job, assume that the fee is $50.00.

No factory-mounting charges are involved in this sample estimate because your crew will mount the relays within the heat pump room enclosures, and you have already allowed for that labor on the space rough-in estimate sheet.

In this sample job, there is no need for a subcontract since your own electricians will mount and wire the devices. However, you could have subcontracted the work to another electrical contractor, if you preferred to broker the installation. In that case, you would enter as a peripheral cost the quotation you received from the subcontractor.

As you purchase data sheets for use in submittal, you will develop a feel for the cost of submittal literature. In some cases all you might do is pull master data sheets from a file on control devices and merely run copies. Be sure to enter such costs as you incur, either duplicating costs or outright purchase costs. You might prefer to enter an allowance for submittal literature derived from a percentage figure based on the size of the job. Or you might feel that the cost is part of your overhead. Independent contractors are required to obtain necessary submittal literature for use in the approval process, and, personally, I feel it's part of the job cost. Remember, there will be many jobs that you negotiate which may not have to be submitted. For purposes of this sample estimate, the schematics for which were developed from plans and specs, the submittal data cost entry is $25.00—1 percent of the total cost of devices and materials.

The project specifications generally call for a warranty period during which any defects in workmanship or materials are to be corrected at no additional cost to the owner. The normal warranty period is 1 year, and I feel that an allowance of 5 percent of labor and materials costs is sufficient. Total costs for devices, materials, and labor are $5221.00, and 5 percent of this figure is $261.00. This is a conservative allowance. But good installation practices, coupled with good setup and checkout, make for a successful job.

No outside purchase assistance is anticipated for commissioning the control system estimated here, and so no allowance is entered for a technical specialist.

The spec usually allocates a certain amount of time for owner orientation. During this time, you (or one of your employees) must explain and demonstrate the system to the owner or the owner's representative, so that once the system is turned over, the responsible parties can make adjustments and maintain the system. What frequently happens is that you wind up doing this twice. The architect and engineer make an inspection, and you probably explain the system then. Then, later, you must go over the system again for the benefit of the owner. If possible, you can kill two birds with one stone by combining the owner orientation with the final inspection. Unless the specifications stipulate a stated period for owner training, I think a day of someone's time would be sufficient for each occurrence: Inspection and, later, orientation. To cover this time, I recommend an entry of $150.00 per day for as many days as are necessary. In this case, assume you need 2 days—a cost of $300.00—for inspection and orientation.

The total peripheral costs, as shown in Fig. 15-3, come to $636.00 (entries have been rounded off to eliminate cents). You are now ready to enter the final phase of pricing the job, and that is totaling materials, labor, and peripheral costs and applying overhead and profit so that you can make a marketing decision on the bidding of the job.

Job Cost, Overhead, and Profit

Totaling the far right-hand column in the bid summary sheet (Fig. 15-3) yields a projected total job raw cost of $5857.00. This is what you can expect to spend in order to install the job and see it through the warranty period. Your overhead rate is a rate you must choose for yourself; here 15 percent is used, and this gives an overhead dollar entry of $879.00. Job cost plus overhead subtotals to $6736.00. You must choose your own profit percentage also, but 10 percent is figured here. This produces an anticipated dollar profit of $674.00. Adding in the profit dollars, you arrive at a target bid price of $7410.00. This is approximately where the right price for this job should be, assuming you have covered everything that will be required of you. At bid time, some jockeying of bid prices may occur, depending on how small a margin you need to keep your crew busy. When there are a significant number of control bidders, prices at bid time for this job could range from $6000.00 to $7500.00. Knowing what your costs are is the most important ingredient in planning your bidding strategy. You will have to live or die by the estimate sheet if you win the bid. First, know what your costs are and then evaluate your need for work. Consider your potential customers, and bid accordingly.

From dealing with various customers, you know that some, more than others, can cause you extra expense. Some customers may lean heavily on you for field support because their jobsite personnel are not as experienced as they should be. Some customers may use you to do their troubleshooting during the warranty period. You know who they are, and you might want to keep their various supervisory needs in mind when you bid their work. When you consider your customers, remember how dependent or independent they are.

ESTIMATING GUIDELINES

It should be obvious by now that the bid summary sheet can be only as good as the data that are transferred from the supporting documents. While you can't foresee the ultimate outcome of the project in terms of final costs, you can plan for construction expenditures; in fact, that is really what you are doing when you prepare an estimate—planning your expenditures. As with anything you undertake, when you have a plan, you have a much better chance of succeeding, particularly if you can implement your plan once you have won the bid. Once you have the project, discuss the various segments of the estimate in a joint meeting with your job installation team. Use the estimate when preparing definitive bills of material for purchasing materials, and make sure that the estimate is referred to during engineering so that the work can be implemented as it was estimated—unless through careful engineering the costs can be lowered.

An estimate is just that—an estimate! Accuracy is important, and a thorough knowledge of the project's requirements is necessary so that you can estimate enough labor-hours and materials to accomplish the work in the most economical manner while still meeting the specifications.

An estimate is not supposed to be detailed enough to submit through channels for approval, nor is it expected to serve as a definitive bill of materials. There simply isn't time to account for every nut and bolt required to do a job. But there is a need to establish a method that will ensure sufficient dollars in the right places on the estimate sheet to get the job done. This is why unitized prices are used on the equipment room labor and materials estimate sheet. As the devices required to do the job in the equipment rooms are listed onto the sheet, dollar costs and labor-hour requirements applicable to each device are entered as a unit. The same is true with tubing and with EMT and wire. Costs per 100 ft are unitized to include wire; you need to develop these costs for yourself, and to keep them current, so that correct figures can be plugged in. Many electric devices will require just two wires, but when three or more are necessary for any given control, you must increase the cost per 100 ft to reflect however many wires are involved. The same holds true for pneumatic devices. If more than two tubes are needed, increasing the cost per 100 ft will allow you to cover the additional tubing.

Unitizing is also used in Table 3-1, where labor-hour units are listed for mounting, connecting, calibrating, setting up, and running 30 ft of interconnecting EMT with two wires, or two tubes as required for a pneumatic control. Labor-hours relating to the running of tubes and wires are determined on the basis of 3/4-in EMT. Unitizing is very practical at the time that the estimate is being prepared, and, as has already been pointed out, produces an estimate that predicts with sufficient accuracy the rough-in materials costs, even though the mix in terms of size and wires will be different.

The consistent use of the estimating method in this text will permit you to develop a feel for where the bottom line leaves you. That is to say, while knowing what your costs will be is important, you can best evaluate how valid those costs are for you after you examine the costs of completed jobs and compare them with your estimated costs. This way, you can look at shop efficiency, project management competency, and employee motivation in light of past jobs and apply appropriate factoring to make the method in this text your personal method. I feel the technique of estimating offered here is a good way to arrive at projected controls installation costs, and it will work even better for you when you add your own experience.

ESTIMATING A PNEUMATIC CONTROL SYSTEM

Let us now look at an estimating exercise based on a pneumatic control system, which is to be installed in a single-story building. In this example, one large air handler will serve 45 VAV boxes out in the occupied spaces. Low-pressure steam at 5 lb will be used for preheat and also for reheat coils downstream from the VAV boxes. Additionally, this steam will be utilized in a heat exchanger for generating hot water for perimeter radiation to offset skin loss in cold weather.

A water chiller will serve the air handler when mechanical cooling is required. Figure 15-4 shows sketches that were developed to assist in the takeoff from the following sequence of operations.

Air-Handling Unit Control

> The unit shall be started and stopped by a morning start-up optimizer, which, during the heating season, shall delay bringing the supply air and return air

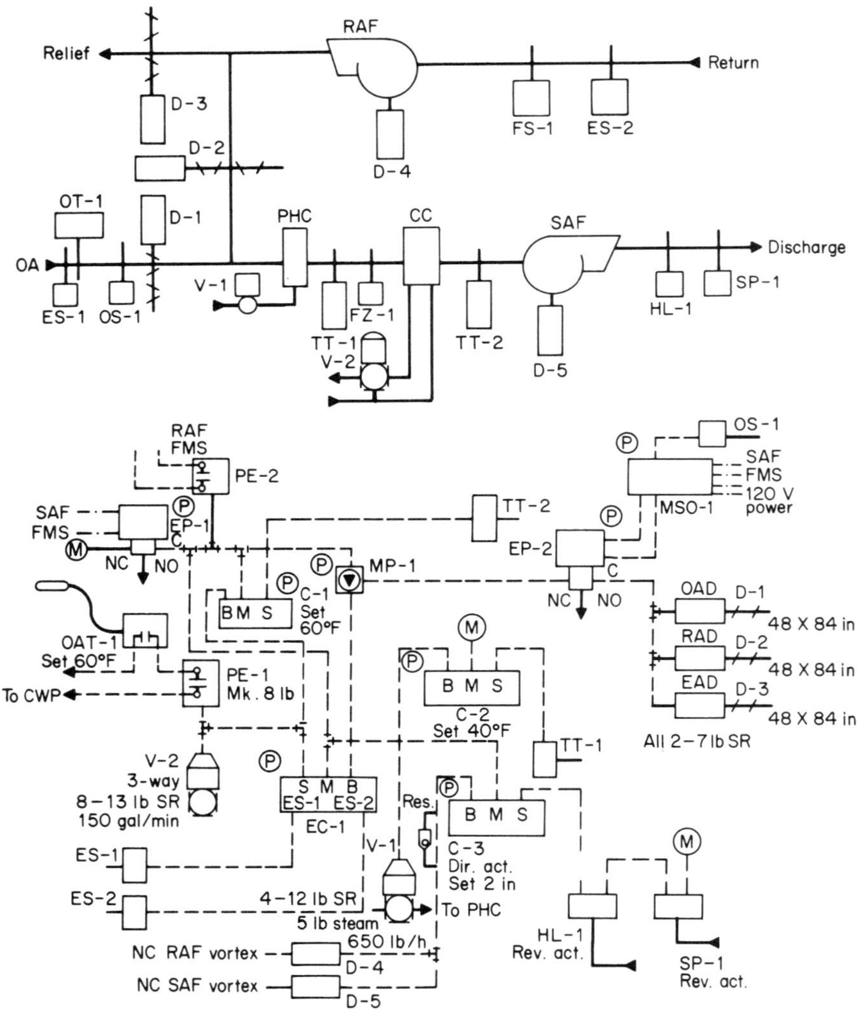

Figure 15-4 Air-handling unit controls.

fans on line until the latest possible time necessary to reinstate occupied daytime temperatures in the building. As the weather grows colder, start-up shall automatically be advanced to an earlier time on the basis of the outside air temperature and the greater heat loss from the building occurring at any given outside air temperature. The introduction of outside air shall not occur until it is time for personnel to begin arriving. The return air fan shall be interlocked to start anytime the supply air fan is running.

Mixed air temperature sensor TT-1 shall provide continuous monitoring of the mixed air approaching the cooling coil so that a minimum mixed air temperature of 40°F can be maintained through the modulation of the preheat coil steam control valve.

Coil discharge temperature sensor TT-2, located to sense air temperatures leaving the cooling coil, shall provide continuous monitoring of the air being delivered to the occupied spaces so that a discharge air temperature of 60°F can be maintained through the sequencing of automatic outside air and return air dampers and the chilled water coil control valve. On a rise in temperature above 60°F, the outside air and relief air dampers shall be modulated open as the return air damper is modulated closed, provided that an enthalpy controller senses that the outside air is a lesser cooling load than the return air, and provided that the morning start-up optimizer has restored daytime temperatures and is permitting such introduction of outside air. Modulation of the dampers shall be progressive and shall proportionately go beyond the minimum outside air setting predetermined by adjustment of a minimum position selector switch. Once the unit is on full outside air, if the demand for cooling continues to increase as sensed by TT-2, the chilled water pump shall be started, and the cooling coil control valve shall be modulated for a greater flow of chilled water through the coil, except that the chilled water pump shall not be permitted to start unless the outside air temperature is 60°F or higher.

When the unit is in operation, static pressure sensor SP-1 shall monitor the duct static pressure occurring two-thirds of the distance down the longest run of duct from the unit, so that a desired water column static pressure of 2 in can be maintained for proper operation of the VAV boxes serving the occupied spaces. As the duct static pressure rises at the sensor, the normally closed supply air and return air fan inlet vanes shall be modulated toward the closed position to reduce the output of the fans and effect a lowering of duct static pressure.

At night during the heating season, when the morning start-up optimizer has the unit off, a night thermostat shall be able to restart the fans, cycling them on and off as required to maintain a minimum temperature of 55°F in the building. During such nighttime cycling of the unit, the space thermostats shall have their reheat coil valves open so that cycling of the unit fans can accomplish space heating. No fresh air shall be drawn in.

Should the return air temperature rise to 135°F or should the mixed air temperature fall to 35°F, both the supply air and return air fans shall stop, the outside air dampers shall close, the fan inlet vanes shall close, and the steam preheat coil control valve shall remain under control. Additionally, the same process shall occur whenever the fans are stopped for any reason.

The morning start-up optimizer shall feature manual override capability so that, should it be desired to use the building during periods when the optimizer has the unit down, normal occupied operating conditions can be intentionally restored.

Chilled Water and Condenser Water Control

When the chilled water pump is called upon to run, the condenser water pump and the water chiller shall start through interlock. At the same time, the cooling tower fan shall be enabled through interlock with the condenser water pump starter. The water chiller shall operate from its self-contained capacity controls, but the cooling tower fan shall be controlled by the temperatures of the condenser water leaving the tower in sequence with a varying volume of water flow over the tower. Condenser water temperature sensor TT-3 shall continuously monitor the temperature so as to permit the sequence described. As the temperature of the water rises at TT-3, a condenser water control valve shall be modulated to deliver a greater volume of water over the tower while simultaneously bypassing less volume. Once the control valve is positioned for full tower flow, a continued rise in the condenser water temperature shall cause the cooling tower fan to start.

Space Temperature Control

Space thermostat ST-1 shall control its respective VAV box or boxes. On a rise in temperature, the reheat coil control valve shall first be closed gradually. Once the reheat valve is fully closed, a continued rise in the space temperature at the thermostat shall cause the box to deliver increasingly greater quantities of 60°F air to the space under control. Between the closing of the reheat coil control valve and the opening of the damper within the box for a greater flow of air, there shall be a deadband of 5°F during which time the box is at minimum in terms of air flow and the reheat coil control valve is closed.

Perimeter Radiation Control

On a fall in the outside air temperature below 60°F, the perimeter radiation hot water pump shall start. Once the pump has started, the hot water heat exchanger shall be placed under control. Perimeter hot water radiation temperature sensor TT-4 shall continuously monitor the temperature of the water being delivered to the radiation so that a discharge temperature to the radiation which is reset inversely with the outside air temperature can be effected. When the outside air temperature is 10°F, the hot water temperature shall be 190°F. As the outside air temperature rises to 55°, the hot water temperature shall be gradually and inversely reset to 100°F. Such reset shall be inversely proportional and linear. Should the outside air temperature rise to 60°F, the hot water pump shall be stopped and the heat exchanger steam control valve shall be closed. This steam control valve shall close should the hot water pump be stopped for any reason. Continuous readout of the hot water temperature and the outside air temperature shall be provided on the face of the local control panel.

Compressed Air Supply Control

A 3/4-hp air compressor with a 208-V three-phase electric motor mounted on a 30-gal receiver shall be provided. The compressor shall be complete with an oil-level dipstick, a pressure switch for control of the pump motor, a belt guard, a high-pressure relief valve, a tank pressure gauge, an automatic tank drain, and all appurtenances necessary for a completely automatic operating unit.

In addition, a refrigeration-type compressed air dryer with a capacity of at least 15 cfm, suitable for operation at 120 V ac with a cord and plug, shall be provided. The unit shall have an automatic trap and a high-temperature alarm light. The necessary filter and pressure regulator and low-pressure relief valve to ensure clean air at the proper pressures for operation of the pneumatic control system shall be furnished.

General

The sequences of operations given for a rise in temperature shall occur in reverse on a fall in temperature.

The control dampers shall be provided by the controls contractor but installed by the sheet metal contractor under the controls contractor's supervision. Control damper sizes shall be shown on the plans, but shall be submitted to the sheet metal contractor for verification of sizes and for approval before ordering. The control dampers shall permit not more than 5 percent leakage at 4 in of water column static pressure. The dampers shall be of the proportional opposed-blade type with blade edging and end seals. The dampers shall operate smoothly over their entire range, and the pneumatic actuators shall be sized large enough to permit satisfactory operation.

Steam control valves shall be sized for not more than a 2-lb pressure drop through the valves when the valves are fully open, and shall be designed specifically for low-pressure steam service. Chilled water valves shall also be sized for not more than a 2-lb drop through the valves when the valves are fully open to the chilled water coil. Control valves 2 in and smaller shall have screwed ends, while valves 2 1/2-in and larger shall have flanged ends.

Pneumatic tubing shall be instrument-grade polyethylene flame-retardant tubing where installed in concealed, accessible locations, such as above lift-out ceilings and in accessible chases. Such tubing in these concealed locations may be run exposed, but in a workmanlike manner, properly supported and protected from abrasion and high temperatures. Within equipment rooms, and where run exposed, polyethylene tubing shall be run in EMT, or hard-drawn copper may be used.

System Operation

The sketches in Figs. 15-4 to 15-9 have been developed from the sequence of operations and information given in this specification, which is basically a performance specification. In Figs. 15-4 and 15-8, the devices identified by a P within a circle are to be mounted in or on a control panel, and so for this exercise the control panel estimate sheet will be used. Damper sizes are given as if they had already been determined from the mechanicals.

Start-Up The specification calls for optimization of unit start-up time and then activation of the unit's ventilation cycle when occupancy begins. The control circuit for the supply air fan is closed by the morning start-up optimizer MSO-1 as determined by programming and outdoor temperature input, and possibly, in some cases, indoor temperature input. EP-2 is controlled by the vent cycle contact, which is indexed by the actual time of day, preset to coincide with the arrival of people. Interlock between the supply air and return air fans is accomplished pneumatically, because EP-1 is energized as a result of its parallel connection with the motor starter holding coil. To prove pull-in of the starter mechanism, the EP is also connected in series through an auxiliary contact within the motor starter. From Figs. 15-5 and 15-6 you can see that the plans envision the use of combination disconnects and motor starters locally situated to serve and protect the motors they control. Thus, you can conclude that a motor control center is not involved. You can further surmise that the

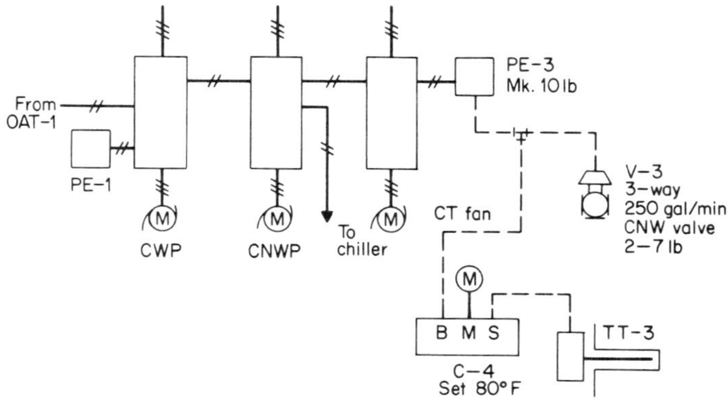

Figure 15-5 Chilled water and condenser water controls.

combination starters for the two fans are not side by side. In this case, the emphasis is on minimizing the control wiring requirements, and so a choice has been made to run a pneumatic line to accomplish the interlock instead of hard-wiring the two starters together. The approach could have been purely electric. This is typical of decisions that you must make when the control wiring is your responsibility, which it is for this sample job.

Discharge Temperature The next item in the sequence of operations is the control of the unit's discharge temperature. Controller C-1, the controller for the discharge temperature, is located in the local control panel, which is quite typical. The branch line, or output, from C-1 goes to three places: to PE-1, to control valve V-2, and to enthalpy controller EC-1. The receiver controller-sensor approach is extremely popular, since grouping interacting devices locally in a panel tends to cut down on equipment room tubing. C-1 is direct-acting; and so as the linear, direct-acting signal from TT-2

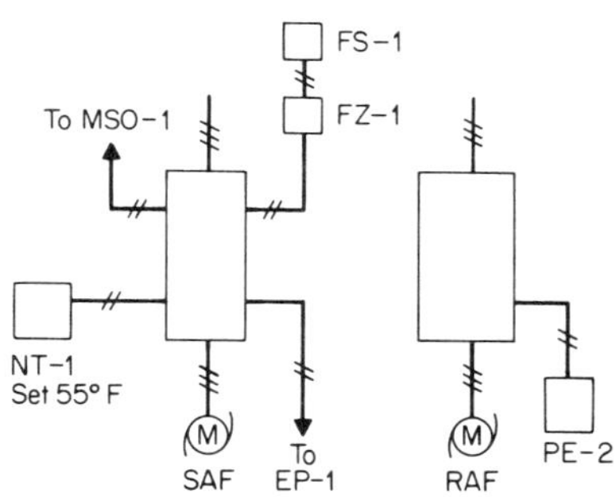

Figure 15-6 Air-handling unit supply air fan and return air fan controls

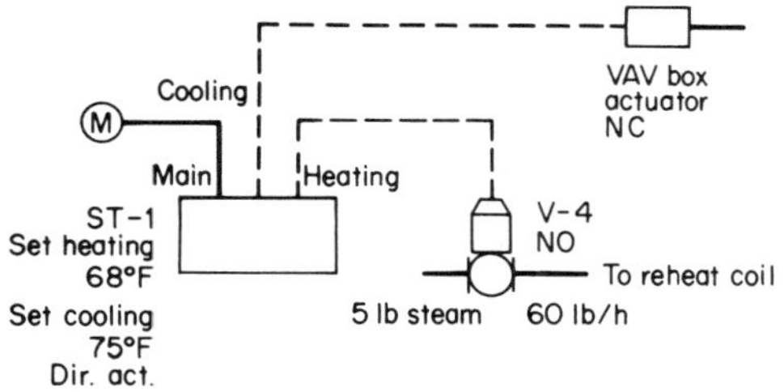

Figure 15-7 Space temperature controls VAV boxes with steam reheat.

rises, indicating a rise in the discharge temperature, the output, or branch line, from C-1 increases also, as the signal from TT-2 crosses its setpoint and throttling range. Notice the spring ranges and set point of PE-1. The pneumatic motors for the three dampers, D-1, D-2, and D-3, feature an "early" spring range of 2 to 7 lb, and they will react first. On a continuing rise, the damper motors will be fully stroked. Then PE-1 will make its electric contact and close its electric circuit at 8 lb. This starts the chilled water pump and associated water chilling equipment to prepare the unit to deliver cooling from mechanical sources. Eventually, as the discharge temperature continues to rise, control valve V-2, with a spring range of 8 to 13 lb, gradually increases the volume of chilled water reaching the cooling coil within the air handler.

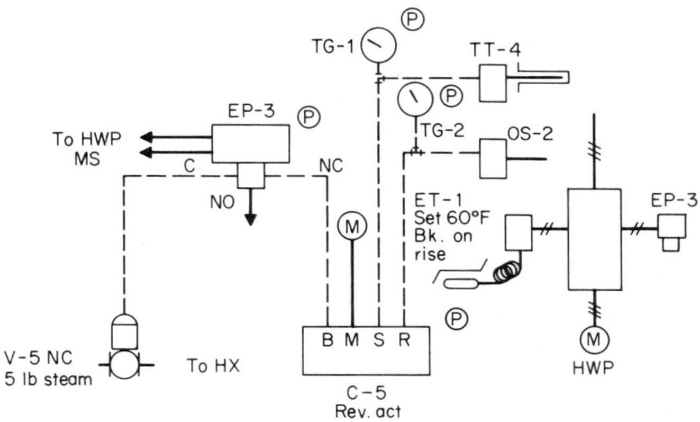

OA	HW
10°F	190°F
55°F	100°F

Figure 15-8 Perimeter radiation controls.

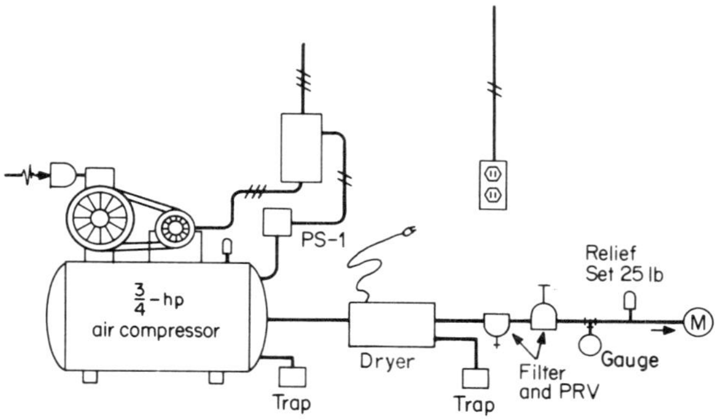

Figure 15-9 Compressed air supply.

At this point, some conditions are imposed by the sequence of operations that must be allowed for. First, the dampers may operate only if the outside air represents a more economical load to the cooling coil than the inside air does. In a situation where the outside air damper is open and the chilled water valve has flow through the coil, the lesser load, in terms of cooling, will come from the air that has the lowest total heat to be removed. Determining which has the lowest total heat is the job of EC-1, the enthalpy controller. With an enthalpy sensor in both the outside and return air, EC-1 can make such a determination. When the outside air does not represent the lighter load in terms of cooling, operation of the ventilation cycle is inhibited.

Secondly, when nobody is in the building at night and thus no fresh air is needed, the outside air dampers should not open. EP-2 ensures that they stay closed by blocking the signal to the damper motors while exhausting pressure from the line going to them. Modulation of the dampers, when conditions are right, overcalls a preset minimum position of the dampers for greater amounts of fresh air. Minimum position switch MP-1 makes sure that a predetermined amount of fresh air is being drawn in at all times to forestall the buildup of stagnant air when people are in the building.

Finally, with respect to cooling, an electric remote-bulb thermostat wired in series with PE-1 prevents the water-chilling equipment from operating unless the outside air temperature is 60°F or higher. When the outside air temperature falls below 60°F, it should be possible to get a discharge air temperature of 60°F with the introduction of 100 percent outside air.

Static Pressure Static pressure is the next item addressed in the control scheme. Notice that the spec calls for positioning static pressure sensor SP-1 two-thirds of the way down the longest run of ductwork. This is to ensure that sufficient pressure exists at the remote areas of the building during operation of the unit. The tubing should appear on the space rough-in estimate sheet.

SP-1 is not a fan room device! However, the high-limit controller is, for it can be located close to the fan. Static pressure controller C-3 is in the local control panel, and it controls two damper actuators on the inlet vanes to the supply air and return air fans, one actuator per fan. Determine ahead of time who furnishes these actuators. There is no need to include them and run the risk of being too high in your bid when with a little effort you can determine if the fan supplier is providing these actuators with the fans. If you must furnish them, obtain torque requirements from the fan supplier's rep so that you know how large a motor to provide.

The fan vortex inlet vanes are normally closed. That is, in the absence of air pressure to the vane actuators, the inlet vanes will be closed. This is to ensure that start-up of the fans will occur without any load on the fan. And a restrictor is in the line from the branch output of C-3 to the vortex actuators to provide for slow opening, thus allowing sufficient time for the buildup of static at the far ends of the duct runs so that SP-1 can sense the buildup and report to C-3. Such an arrangement also helps to prevent the static pressure controller from hunting. Some of these precautionary measures have also been noted in earlier portions of this book. But they are worth repeating because they are important to keep in mind as you take the job off to ensure your competitive posture and the workability of the system.

To guarantee that the inlet vortexes will return to the closed position when the fan is off, EP-1 air is used as the "main" to C-3; when the fan stops, the main air to C-3 is withdrawn, and actuators D-4 and D-5 spring-return the vanes closed. Note that SP-1 is reverse-acting, so that on a fall in static pressure, its output builds. In this way, in-line high-limit controller HL-1 can overcall SP-1 to keep the signal from reaching C-3 and thus prevent overpressurization of the unit.

During the winter, the unit will be off at night. But the spec calls for restarting of the unit should the temperature in the building fall to 55°F. By wiring night thermostat NT-1 parallel in the starter to the fan contact in MSO-1, the fan can be brought back on when necessary. No outdoor air will be introduced, because EP-2 prevents the dampers from even assuming their minimum position. And the unit will deliver heat, even though there is no heating coil within the unit. The space thermostats controlling the VAV boxes and reheat coils are set for 68°F, and so they will be "cold," their box actuators will be positioned for minimum cool air flow, and the reheat valves will be open.

Temperature Extremes Next, the spec addresses temperature extremes within the unit that would indicate an emergency condition and/or damage the unit. Electric firestat FS-1 and electric freezestat FZ-1 are wired in series with the fan's holding coil circuit of the motor starter. The electric contacts of FS-1 and FZ-1 are normally closed. A rise in temperature at FS-1 to 135°F or a fall in temperature at FZ-1 to 35°F causes each to open its contact, and this stops the fan. Of course, EP-1 drops out, and so C-1 loses "main," as does

C-3. But straight main is piped to C-2, and so it can continue to maintain a temperature of 40°F should the fan go down in the winter.

The morning start-up optimizer should have an override feature built in, so that building management can restore normal daytime conditions when occupants are working late. In some specs, locally accessible override timer switches are called for, but in this case, we'll assume that an override at MSO-1 is all that's necessary. The override feature, however, should have a recycling capability, subject to indexing for unoccupied conditions as the end of the next day approaches.

Mechanical Cooling System Control of the mechanical cooling system, the next topic in the spec, presents a good mix of electrical and pneumatic work. The condenser water pump motor starter control circuit is wired through an auxiliary contact in the chilled water pump motor starter, and the cooling tower fan motor starter control circuit is wired through an auxiliary contact in the condenser water pump motor starter. The water chiller control circuit is also wired through an auxiliary contact in the condenser water pump motor starter; this prevents operation of the chiller unless both pumps are in operation. Chiller manufacturers usually also require a water flow switch for proof of flow; this is an area of control wiring that bears checking out, particularly if you will have to do it.

At this point, the cooling tower fan motor starter is enabled merely by startup of the condenser water pump. PE-3, which is also connected in series with the cooling tower fan starter coil circuit, is "made" by condenser water temperature controller C-4 on a rise in condenser water temperature above 80°F. However, the fan won't start until control valve V-3 is delivering all the water over the tower. This requirement can be met by setting the PE for 10 lb, which is above the spring range of V-3.

These approaches are fairly standard in the control industry, but inventiveness is always in order when you can see a more economical way to accomplish the same thing. Some control applications allow more latitude in the exercise of ingenuity, particularly as this applies to air handlers.

Space Temperature Space temperature is maintained by thermostats in multiple locations throughout the building, as shown in the mechanical drawings. The locations of the thermostats, plus the locations of the boxes and reheat coils, provide targets for the tubing runs. These devices are listed on the space rough-in estimate sheet. The spec states that a rise in temperature will cause the steam control valve at the reheat coil to close and that a continuing rise in temperature will cause the internal damper of the VAV box to open for a greater flow of cool air into the space. Often the pneumatic damper actuators will come premounted on the box. A flexible polyethylene tube is usually stubbed out for connection by the controls contractor, or a pneumatic fitting is identified as the point for connection of a branch line from the thermostat. When the damper actuator is furnished with the box, determine what the normal state of the damper is and verify what type of thermostat is needed. For purposes of the exercise here, we'll assume that the box damper actuator holds the internal damper in its minimum position in the absence of control air and, further, that the steam coil reheat control valve is normally open to steam flow in the absence of control air. Spring ranges do not have to be split between the two controlled devices, because two separate signals from the space thermostat are piped to their respective device and a deadband in between operation of the two is called for. An energy-efficient thermostat should be selected when a null or dwell period between delivery of the two signals is provided for.

Watch your tubing. Two lines run from the thermostat to the controlled devices. This is an excellent opportunity to use bundled two-tube polyethylene, because the specs allow polyethylene above accessible ceilings.

Perimeter Radiation The next item in the sequence of operations deals with skin loss in cold weather. As the outside air temperature falls, skin loss can be offset by inversely resetting the hot water that flows through finned-tube perimeter radiation. This can be accomplished by using pneumatic receiver controller C-5, which is piloted by a sensor measuring the temperature of the hot water flowing to the radiation and by another sensor measuring the outside air temperature. The branch line from C-5 is taken through EP-3 so that when the hot water pump is stopped, air can be exhausted from heat exchanger steam control valve V-5, allowing it to assume its normally closed position. The reset schedule is usually specified, and this is a setup function. This information will appear on the shop drawings to the field. Above 60°F outside air temperature, the hot water pump is stopped by electric thermostat ET-1, whose remote bulb is located in the outside air under a sunshield. Readout of both the hot water temperature and the outside air temperature is provided for with pneumatic receiver gauges on the local control panel calibrated to match the range of pneumatic temperature sensors TT-4 and OS-2.

Compressed Air Supply Finally, the spec discusses the source of compressed air, and the compressed air equipment is diagramed along with the control schematics to ensure that it will be listed when the takeoff is made from these schematics.

As the takeoff is made and the sketches are developed, the requirements for two or more of the same scheme can be indicated by noting right on each sketch the number of schemes the sketch is typical for. For example, you would note that the sketch for space temperature controls for the VAV boxes and reheat coils is typical for 45.

Control damper sizes are listed on the sketch too. The dampers are selected on the basis of the requirements for effective sealoff, type of construction, blade configuration, and size.

Control valves are sized on the basis of an allowable pressure drop through the valve of 2 lb when the valve is passing full flow. Screwed ends and flanged ends are determined by the size of the valve, according to the spec.

The spec provides the option of using either polyethylene and EMT or hard-drawn copper in equipment rooms and other exposed work. In a mechanical room with one line, it is usually more economical to run a hard-drawn copper tube rather than use EMT and pull in one polyethylene tube. Above suspended ceilings and in accessible locations, such as chases, polyethylene, properly installed, can be used—but be sure it is rated fire retardant.

Using the Estimate Sheets

The next step is to put this job onto the estimate sheets (see Figs. 15-10 to 15-13 later in this chapter) and begin to price it. Prices for the control devices and tubing have been assumed. Since these prices can change, you *must keep your list of device and materials prices up to date* so that your estimate is based on current data. Your estimate will be only as good as your takeoff and the prices shown on your estimate sheets. Since this sample job has no mechanical drawings from which to determine distances on the floor plans, some assumptions have been made here too.

The control diagrams cover the air handler first, and so you will use the equipment room labor and materials estimate sheet (Fig. 15-10) first. On this job, the air handler, the motor starters you are involved with, the heat exchanger, and the air compressor are all in the same equipment room, and so all your controls are covered on the equipment room labor and materials estimate sheet. As you begin, remember that controls within the control panel are not to be entered on this sheet. A common control panel for all panel-mounted equipment room controls will be used, and those items within the panel will be considered later on the control panel estimate sheet (Fig. 15-12).

PEs Three PEs are the first items listed on the equipment room labor and materials estimate sheet. Each is to be located outside the control panel close to the motor starter it will control. PE-1 will be wired to bring on the chilled water pump; PE-2 will be wired to serve as the interlock with the supply air fan by bringing on the return air fan; and PE-3 will start the cooling tower fan in response to the branch line from C-4. A cost of $18.00 has been allowed for each PE. The number of labor-hours allowed for each PE—3.5—has been taken from Table 3-1, the table in Chap. 3 which lists equipment room labor-hours; the labor-hours for each control on this estimate sheet have all been taken from Table 3-1.

Temperature Sensors The estimate sheet shows that six temperature sensors are required: four for sensing air temperature and two for sensing water temperature. TT-1 checks on the air temperature leaving the preheat coil. TT-2 is the discharge temperature sensor and checks on the air temperature leaving the unit. OS-1 monitors the outside air temperature on behalf of the morning start-up optimizer, while OS-2 performs the same service for C-5, the hot water temperature controller. Quite frequently, different outside air tem-

Equipment Room Labor and Materials Estimate Sheet

Job name ABC Building Date ________
Estimator Harry Edwards Page 1 of 4
Room or unit number Equipment room Floor or area ________

Control	Pneu.	Elect.	Qty.	Cost/ea., $	Total cost, $	Labor/ea., h	Total labor, h	A	B	C	D	E
PE	X	X	3	18	54	3.5	10.5	3		3		
Air temp. sensor	X		4	55	220	4.0	16.0			4		
H_2O temp. sensor	X		2	50	100	3.5	7.0			2		
Remote bulb stat		X	2	60	120	4.0	8.0	2				
Damper motors	X		5	150	750	4.0	20.0			5		
Static pressure	X		1	85	85	5.0	5.0			1		
Enthalpy sensor	X		2	225	450	3.5	7.0			2		
3-in 3-way CW valve	X		1	340	340	4.0	4.0			1		
2-in 2-way steam valve	X		1	185	185	4.0	4.0			1		
4-in 3-way CWW valve	X		1	450	450	4.0	4.0			1		
Fire stat		X	1	15	15	2.5	2.5	1				
Freezestat		X	1	45	45	4.0	4.0	1				
Filter and PRV	X		1	100	100	5.0	5.0			1		
Nightstat		X	1	25	25	2.5	2.5	1				
Panel rings	—	—	2	50	100	One	2.0	—	—	—	—	—
Total	21	8	28		3039		102	8		21		

Transfer total equip. room control device costs $ 3039 to bid summary sheet, line 2

Transfer total equip. room hours 102 (X 1.3)* = 102 to bid summary sheet line 11

(A) Elect. controls wired in EMT: 8 X 30 ft/ea. = 240 ft
240 ft EMT and wire (X 1.3)* = 240 ft X $ 39.00 /100 ft = EMT and wire $ 94

(B) Elect. controls wired rigid: ____ X 30 ft/ea. = ____ ft
____ ft rigid and wire (X 1.3)* = ____ ft X $ ____ /100 ft = rigid and wire $ ____

(C) Pneu. controls piped in EMT: 21 X 30 ft/ea. = 630 ft
630 ft EMT & tubes (X 1.3)* = 630 ft X $ 36.00 /100 ft = EMT and tubes $ 227

(D) Pneu. controls tube bundled: ____ X 30 ft/ea. = ____ ft
____ ft tube bundled (X 1.3)* = ____ ft X $ ____ /100 ft = tube bundled $ ____

(E) Pneu. controls tube copper: ____ X 30 ft/ea. = ____ ft
____ ft tube copper (X 1.3)* = ____ ft X $ ____ /100 ft = tube copper $ ____

Transfer total installation material costs to bid summary sheet line 7 = $ 321

* Use factor only if area over 750 ft^2

Figure 15-10 Equipment room labor and materials estimating exercise no. 2.

perature sensors are required for an optimizer and a receiver controller, due sometimes to the special temperature ranges required for the morning start-up optimizer. These two types of air temperature sensors are generally priced the same, however. Most manufacturers will send bulb support brackets along with their airstream sensors when remote-bulb sensors are purchased, and usually manufacturers will package capillary support clips with averaging elements. Both rigid-stem and remote-bulb sensors may feature stick-on templates for drilling when the device is to be mounted. For these air temperature sensors, hours for capillary-type or remote-bulb sensors are selected from Table 3-1, and the 4.0 allowed for each one is multiplied by 4, for a total of 16 h for the four sensors.

TT-3, one of the two water temperature sensors, is located in the condenser water coming off the cooling tower back to the water chiller. TT-4, the other water temperature sensor, is located in the hot water leaving the heating hot water heat exchanger. The temperature range chosen for the condenser water might be 0 to 100°F, while the temperature range selected for the heating hot water might be 40 to 240°F. Both TT-3 and TT-4 should be priced the same, even though their temperature ranges may be different, and both should come packaged with separable socket wells. Even if they're not packaged with separable wells, price them that way so that your price on the estimate sheet includes money for the devices and their necessary accessories. Generally, controls and accessories are listed separately on manufacturers' price sheets. The costs of controls and accessories should be combined in order to arrive at a unit price that covers all hardware necessary for installation. The two water temperature sensors are rigid-stem devices, and 3.5 h, taken from Table 3-1, has been allowed for each of them.

As you enter each unit on the estimate sheet, systematically check whether the piece is pneumatic only, electric only, or both pneumatic and electric. This proves helpful in reminding you what to enter in Columns A to E, which determine counts for the in-room allowances of rough-in material.

Remote-Bulb Thermostats Two electric remote-bulb thermostats are listed next on the equipment room labor and materials estimate sheet. The first, OAT-1, is located with its bulb in the outside air and acts to lock out operation of the chilled water pump below 60°F. The second, ET-1, is located in the outside air also and acts to lock out operation of the heating hot water pump above 60°F. One double-pole double-throw thermostat could have been used to accomplish both functions, or one thermostat and two relays could have been used. However, neither would provide enough flexibility when the system is in operation. Different set points might be required so that some overlap can be introduced when the interior spaces need mechanical cooling (as a result of heavy interior occupancy loads) and the perimeter (with a light occupancy load and heavy fenestration) needs a touch of perimeter radiation heating. By using two thermostats, this can be done. Since the thermostats are electric instruments and the wiring system will be EMT, 4.0 h each is selected.

Damper Motors Damper motors are listed next. Figures 15-4 to 15-9, the takeoff sketches, show that three are required for the automatic dampers on the suction side of the supply fan: one in the outside air, one in the return air, and one in the relief air of the return air fan. For this sample job, it is assumed that you must also provide the damper motors for the vortex inlet vanes on the supply air and return air fans. As you price damper motors during the course of your estimating, you will have to determine how large they need to be. The size of the damper and the torque required to operate each damper must be taken into consideration as you look at the various sizes of damper motors. Manufacturers furnish actuator capability information with their data sheets and/or in their catalogs. For the sample job here, five damper motors of the same size have been selected at a cost of $150.00 each. A separate linkage kit may need to be ordered, but the damper motor price *should* include the cost of damper motor accessories; if it doesn't, list the accessories separately.

Static Pressure Devices Static pressure control is an important ingredient of an air-handling unit's operation when VAV terminal units are employed. The output of the fans must be reduced as boxes close off to primary air to prevent overpressurization of the ductwork. Two static pressure devices are required here. One is a static pressure safety, and the other is a static pressure sensor. Both are field-mounted. But the static pressure sensor is really a space-mounted control. The high-limit, or safety controller (HL-1 in Fig. 15-4), will be mounted in the equipment room on the ductwork of the air handler in the leaving air. It will sense an overpressurization condition occurring at the unit, the first place overpressurization will occur. The static pressure sensor (SP-1 in Fig. 15-4) needs to be positioned out in space above the ceiling in the conditioned areas, mounted on the side of the ductwork, to sense static pressure some distance from the unit. This is necessary to ensure sufficient static pressure at the VAV boxes all along the way, even at the farthest runs of ductwork. Thus, for estimating purposes, SP-1 needs to be entered on the space rough-in estimate sheet. However, HL-1, the static high-limit controller, belongs on the equipment room labor and materials estimate sheet.

Enthalpy Sensors Two enthalpy sensors are also situated in the equipment room; they are mounted in the ductwork at the unit and they sense the total heat content of the outside air and return air. The enthalpy sensors are usually self-contained and do not have a capillary tube to a remote bulb, and so the labor-hours for each should be taken from the "Rigid-stem" row, rather than the "Remote-bulb" row, of the "EMT pneu." column of Table 3-1. Thus, the selection is for 3.5 h. The enthalpy entry listed farther down in the table is for a field-mounted enthalpy *controller*.

Control Valves The next items listed are the control valves. The larger one at the air handler is a three-way chilled water valve, and as noted on the sketch in Fig. 15-4, it needs to pass 150 gal/min. The general section of the specification governing this sample job calls for a 2-lb pressure drop through the control valve when it is passing full flow. To select the proper-size valve, you need to size for water flow requirements (C_v), which can be found using the formula (from Chap. 2)

$$C_v = \frac{Q_w}{\sqrt{P_d}}$$

where Q_w is the water flow in gallons per minute and P_d is the allowable pressure drop. Dividing 150 gal/min by the square root of the allowable pressure drop, which is 2, produces a C_v of 107 (150 divided by 1.4). From the manufacturer's data, you would select a valve that has that particular coefficient of flow—probably a 3-in valve.

The smaller valve on the air handler must pass 650 lb/h of 5-lb steam. The equation (also from Chap. 2) for finding the C_v for this steam valve is

$$C_v = \frac{Q_s \sqrt{V}}{63.5 \times \sqrt{P_d}}$$

where Q_s = pounds per hour of steam,
V = specific volume of 5-lb steam,
63.5 = a constant,
P_d = allowable pressure drop.

Substituting into the equation, you get 650 × 4.518 divided by 63.5 × 1.4, which reduces to a C_v requirement of 33 (2936 ÷ 89). Usually, a 2-in valve will suffice, and that is what's shown in Fig. 15-10. Be sure that the valve costs include the costs of actuators and linkage, when you are responsible for these items.

The final control valve to be considered is the condenser water valve, and it is to pass 250 gal/min in the wide-open position. Dividing 250 by 1.4 produces a C_v of 178—a 4-in valve will serve this purpose. An allowance of 4 h is allocated to each control valve to allow your installers time for tubing, stroking, and checking the valve. Additionally, your installers will be expected to check on how the valve is being installed by the mechanical contractor in order to prevent it from being piped incorrectly.

Firestat and Freezestat The firestat and freezestat are safety controls and are purely electric. When there is tight jurisdiction on a job, the capillary of the freezestat is normally strung by a pneumatic installer, but an electrical installer could do this just as well.

Air Filter and Pressure-Reducing Valve Next, an air filter and a pressure-reducing valve for the air compressor are shown, and they are priced as a single unit. Quite often, you can purchase combination devices nippled together. The nightstat, mounted just through the wall, is shown here too.

Now you must add the columns and bring the appropriate numbers down to their respective locations on the lower portion of the equipment room labor and materials estimate sheet. Transfer the total equipment room device costs, total equipment room hours, and total installation material costs to the bid summary sheet. In regard to the preprinted entry for panel rings, a cursory appraisal of the devices going into the panel indicates a need for two panel rings, placed side by side as a single unit.

Controls and Tubing in Conditioned Spaces Controls and tubing out in the conditioned spaces are considered next. The space rough-in estimate sheet (Fig. 15-11) indicates two different sizes of main: 700 ft of 1/2-in polyethylene and 1100 ft of 3/8-in polyethylene. Branch lines from the thermostats to the boxes and reheat coils require two-tube bundled polyethylene, and 1300 ft is shown on the space rough-in estimate sheet. The price per foot and the

Space Rough-In Estimate Sheet

Job name: ABC Building (single-story) Date: ______

Estimator: Harry Edwards Page 2 of 4

Area or floor	Feet	Size	Material	Cost/ft, $	Total cost, $	Labor/100 ft, h	Total labor, h
1st floor, main	700	1/2 in	Poly	0.09	63	2	14
1st floor, main	1100	3/8 in	Poly	0.07	77	2	22
1st floor, branch	1300	5/32 in	2 Tube poly	0.11	143	2	26
Total					283		62

Transfer space rough-in material costs to bid summary sheet, line 6

Transfer space rough-in hours (X 1.3 for high work) to bid summary sheet, line 9

Space devices, stats, sensors, etc.

Area or floor	Qty.	Model	Cost/ea., $	Total cost, $	Labor/ea., h	Total labor, h
1st floor	45	ST-1	40	1800	1	45
1st floor	45	1/2 in Valves	65	2925	1	45
1st floor	1	SP-1	85	85	1	1
Total	91			4810		91

Transfer space device controls costs to bid summary sheet, line 1

Transfer space mounting hours X 1.2 difficulty factor,* if appropriate, to bid summary sheet, line 10

*Ladder required

Space device, new work: connect, mount, and calibrate, 1 h each

Space device, old work: connect, mount, and calibrate, 1.2 h each

Figure 15-11 Space rough-in estimating exercise no. 2.

hours per 100 ft are entered on the equipment room labor and materials estimate sheet and then extended to develop column totals. Two-tube bundled polyethylene is figured also for the combination main-and-branch run to static pressure controller SP-1, located two-thirds of the way down the longest run of ductwork. This will be a bare polyethylene installation, in keeping with the spec, which permits its use above accessible ceilings.

The space temperature control devices consist of 45 thermostats and 45 1/2-in control valves for controlling the steam at the steam reheat coils. The capacity requirement is 60 lb/h, and substituting this into the equation for finding the C_v for steam valves gives a C_v of 3. We will assume here that this can be met with 1/2-in control valves. The static pressure controller is listed in the space devices section too and is totaled in to find the total cost and total labor-hours for work in connection with the space devices. Space device dollars and hours are transferred to their respective locations on the bid summary sheet.

Control Panel What about the control panel? The decision has been made to build it in the shop. One morning start-up optimizer is required along with three EPs. Four single-input pneumatic controllers and one dual-input controller are needed. The enthalpy controller, a minimum position switch, and two pneumatic receiver gauges are also on the control panel estimate sheet (Fig. 15-12). The prices have been assumed for the purpose of this sample job, and the labor-hours are taken from Table 10-1, which lists temperature control panel labor-hour units. The morning start-up optimizer consists of a triple electric input device and one pneumatic input device. The three electric inputs are a 120-V ac source, the control circuit from the supply air fan, and another circuit to EP-2. The one pneumatic input is from OS-1. All panel devices, except the gauges, will be inside the panel. A triple-input electric controller is allowed 2.5 h, and a single-input pneumatic controller is allowed 1.2 h. Adding these together equals 3.7 h, which is entered on the control panel estimate sheet. Three EPs are inside the panel and have been allotted 1 h each. The single- and dual-input controllers are pneumatic, and their hour entries are 1.2 and 1.75, respectively. The enthalpy controller is a triple-input pneumatic device (ES-1, ES-2, and C-1), and 2.3 h is allowed. The minimum position switch is a pneumatic switch and is allocated 1/2 h. And two pneumatic gauges on the door are allowed 3/4 h each.

Two subpanels must be allowed for, since there are two panel rings. Two doors are needed too, with two cutouts (two gauges). Allowances for some tubing and wire round out the control panel estimate sheet. The columns must be totaled and the totals transferred to the bid summary sheet. The hours should be multiplied by a factor of 0.8 because of shop fabrication. The panel will not require an air header. Instead, tees will be used as required to pipe main air where it needs to be.

Is a terminal strip needed? Which devices must have electric connection to the field? EP-1 must, and MSO-1 must too. EP-3 will field-connect to the hot

Control Panel Estimate Sheet

Job name ABC Building　　　Date ______

Estimator Harry Edwards　　　Page 3 of 4

Area or room where panel is located ______

Panel devices	Qty.	Cost/ea., $	Total Cost, $	Labor/ea., h	Total labor, h	Pneu.	Elect.
MSO	1	285	285	3.7	3.7	1	1
EP	3	18	54	1.0	3.0	3	3
Single-input contr.	4	160	640	1.2	4.8	4	
Dual-input contr.	1	190	190	1.75	1.75	1	
Enthalpy contr.	1	215	215	2.3	2.3	1	
MP switch	1	35	35	0.5	0.5	1	
Temp. gauge	2	15	30	0.75	1.5	2	
Subpanels	2	20	40	0.5	1.0	—	—
Doors	2	20	40	—	—	—	—
Cutouts	2	—	—	0.25	0.5	—	—
Tubing	130	0.05	6.50	—	—	—	—
Wire	40	0.08	3.20	—	—	—	—
Total	19		1539		19	13	4

Transfer total of panel-mounted device costs (including subpanel, door, wire, and tube) to bid summary sheet, line 3

Enter total panel fabrication time in hours on bid summary sheet, line 12*

For special panels fabrication costs, enter quotes on bid summary sheet, line 8

*For in-shop assembly of panels, hours may be factored by 0.8

Total panel hours 19 X 0.8 = 15

Figure 15-12 Control panel estimating exercise.

water pump starter also, but altogether only eight wires are involved, and so there's no reason not to wire straight in. A terminal strip won't be included. In larger panels with multiple electric terminations, the terminal strip is a big plus. In this panel, however, a shutoff valve is not needed, nor is a fuse block or a PRV. A PRV would be used if high-pressure air were being brought to the panel, reducing the pressure within the panel. No transformers, receptacle, or jacks are involved either. That should do it for the control panel.

Arriving at a Bid Price

Now the three estimate sheets have been prepared, and the information on them can be transferred to the bid summary sheet (Fig. 15-13) for summarization and development of the bid price.

Control System Devices Costs The damper cost and the air compressing outfit costs are entered separately on lines 4 and 5 on the bid summary sheet. I prefer separation of the damper and air compressing outfit costs so that they can be readily identified should negotiation on these items be desired. On line 5, include labor and wire, as applicable.

As you enter the information on your bid summary sheet, you will need to refer to the estimate sheets to obtain the data. For example, the first three lines on the bid summary sheet deal with control devices, and so you will need to obtain information from the space rough-in estimate sheet (Fig. 15-11), the equipment room labor and materials estimate sheet (Fig. 15-10), and the control panel estimate sheet (Fig. 15-12). The devices are listed according to where they will be placed within the building, and the applications engineer who will later develop the working drawings for the job, should you be successful at bidding, will more clearly see what your intent was at the time the estimate was prepared.

The $4810.00 space device costs cover thermostats, control valves, and the static pressure sensor that will be located in the soon-to-be-occupied spaces of the building. The thermostats will be on the walls of offices and other rooms, and the control valves and the static pressure sensor will be above the ceiling.

Equipment room devices amount to $3039.00 in costs and are associated with the air handler and with the equipment used for generating hot water and steam for heating and chilled water for cooling. The filter and pressure-reduction station for the compressed air for the control system will be located within the mechanical equipment room also, and the night thermostat will be located just outside the equipment room, directly through the wall, where it will be placed on a wall of a corridor. In this one-floor building, the equipment room is located in one of the corners of the building and so the placement of the night thermostat just outside of the equipment room wall in the corridor will actually place the night thermostat in a perimeter area for better sensing of falling temperatures when the building is unoccupied. Were the thermostat to be located in the center of the building, the perimeter spaces could become dangerously cold in extremely cold weather before the night thermostat sensed temperatures low enough for it to activate the air handler.

Line 3 on the bid summary sheet shows the costs of the devices in the control panel(s). The subpanel and the door(s) of the panel are costed in at this point also. By the time the costs for the first three lines have been written in, the control devices have been accounted for. In my method for estimating, the control valves are looked upon as control devices, for they probably will be purchased from a major controls manufacturer's catalog along

Bid Summary Sheet

Job name *ABC Building* Date ______

Estimator *Harry Edwards* Bids due ______ Page *4* of *4*

Devices and materials

Control system devices

1. Space devices costs	$ 4810		
2. Equip. room devices costs	$ 3039		
3. Panel devices costs	$ 1539		
4. Damper costs	$ 810		
5. Compressor and dryer costs	$ 3170		
Total devices costs	$ 13368	$ 13368	

Installation materials

6. Space rough-in materials costs	$ 283		
7. Equip. room materials costs	$ 321		
8. Special panel costs	$ -		
Total materials costs	$ 604	$ 604	
Escalation allowance		$ 838	
Sales tax allowance		$ 592	
Total devices and materials costs		$ 15402	$ 15402

Labor

Labor

9. Space rough-in hours	62
10. Space-mounting hours	91
11. Equip. room hours	102
12. Panel fabrication hours	15
13. Travel time hours	32
* 14. Checkout and supervision hours	-
Total labor-hours	302

151 hours @ $ 15 per hour	$ 2265		
151 hours @ $ 8 per hour	$ 1208		
___ hours @ $ ___ per hour	$		
Total installation costs	$ 3473	$ 3473	
Labor travel expenses		$ 480	
Engineering costs		$ 347	
Taxes, insurance and fringe costs on labor		$ 611	
Total labor costs		$ 4911	$ 4911

Related costs

Peripheral costs

Bonds, cleanup, construction shack	$ -	
Permit fees	$ 150	
Factory-mounting charges	$ -	
Subcontracts	$ -	
Submittal data	$ 154	
Warranty	$ 1015	
† Technical specialist's allowance	$ -	
Owner orientation allowance	$ 300	
Total peripheral costs	$ 1619	$ 1619

Total job raw cost		$ 21932
15 % overhead	$ 3290	$ 3290
Job cost and overhead		$ 25222
10 % profit	$ 2522	$ 2522
Job cost, overhead, and profit (bid price)		$ 27744

*Special allowance only, if applicable
†Outside purchased assistance only, if applicable

Figure 15-13 Control systems bid summary estimating exercise no. 2.

with the rest of the control items. Dampers, on the other hand, are frequently purchased from various sources, and their costs appear as the next item on the bid summary sheet. Many mechanical contractors have good connections with suppliers of quality dampers of all types, including control dampers. Routinely, the mechanical contractor buys manual balancing dampers, splitter dampers, smoke isolation dampers, and barometric dampers, and so it comes as no surprise that the manufacturer's representative who sells dampers to the mechanical contractor could probably offer automatic control dampers as well. For this reason, the cost for dampers is entered as a separate cost, so that the mechanical contractor can easily have the option to purchase them from you or from a regular supplier. Indeed, some mechanical contractors have excellent sheet metal shops and may elect to fabricate the control dampers themselves. The $810.00 damper cost includes the cost of three 48 × 84 in control dampers at $270.00 each.

A word of caution about dampers. If the mechanical contractor, and not *you*, is going to supply the dampers, *be sure* that the control dampers that were specified, with edge and jamb seals, are provided and that the mechanical contractor understands exactly what is to be ordered. Remember, the specification for the dampers was in all likelihood in your sections of the specs, and if what is delivered to the job does not meet the standards outlined in specs, it is you who will come under fire. Therefore, participate sufficiently in the ordering of control dampers to ensure that dampers of the quality called for in the specs are ordered. Additionally, make sure the mechanical contractor knows what dampers are covered by your quote. Satisfy yourself that you have taken off all the dampers that you need to be responsible for.

On line 5, the air compressor is accounted for, and along with it, the air dryer. The sum of $3170.00 includes $2200.00 for an air compressor; $600.00 for an air dryer; $250.00 (rounded) for a mechanic and helper for 1 day to set, pipe, and wire them; and another $100.00 for rough-in materials. The price for the compressor must also include accessories for proper operation of the air compressor, such as a motor starter, an alternator (in the case of a duplex unit), a belt guard (partially or totally enclosed, as the spec may require), an oil-level sight gauge (if called for), an automatic trap for removal of condensation from the receiver, an electric motor of the proper voltage, a tank pressure gauge, a relief valve, vibration isolators, an intake silencer, a manual drain cock, ASME certification, an air intake filter, and anything else that is called for. The manufacturers usually supply ASME certification along with the air compressors.

The first five lines are totaled to get the total costs for devices and equipment, in this case $13,368.00. For these devices, installation materials are required in the form of tubing, conduit, and wire. From the space rough-in estimate sheet, $283.00 worth of installation materials is picked up and transferred to line 6. Instrument-grade polyethylene has been estimated, and two sizes of mains are shown, 1/2-in OD and 3/8-in OD. You have done enough air main sizing to know that the first 700 ft out of main needs to be 1/2 in, and from there on

3/8 in will do. Two-tube 5/32-in bundled tubing will serve nicely from the space thermostats to the VAV box and reheat coil. Remember, two branch lines are required from the thermostat out, because two separate signals originate at the thermostat.

Installation Materials Costs The costs of installation materials in the equipment room are now functions of prestated allowances of tube for pneumatic work and wire for electrical work. The equipment room labor and materials estimate sheet is designed for unitized pricing, and the numbers of devices act as the multipliers of given footages for equipment rooms under 750 ft^2. If an equipment room is larger that 750 ft^2, a multiplier of 1.3 is used before dollars per 100 ft are applied. The total installation materials costs ($321.00) are brought from the equipment room labor and materials estimate sheet to line 7 of the bid summary sheet.

There are no special panel costs in this sample job since no assembled panels will be purchased from outside suppliers. The costs for the panel devices, including subpanels and doors, are entered on line 3 of the bid summary sheet, and the cost of the panel rings for the control panel is entered on line 2, because a panel ring is an equipment room item.

Totaling the space rough-in materials costs and the equipment room materials cost results in $604.00 worth of installation materials costs. The escalation allowance in this case is figured at 6 percent of the sum of the total devices costs and the total materials costs ($13,368.00 plus $604.00 multiplied by 6 percent), which equals $838.00. A sales tax of 4 percent is figured on the sum of the total devices costs ($13,368.00), the total materials costs ($604.00), and escalation ($838.00), or 4 percent of $14,810.00, which yields $592.00. Adding together the total devices costs, the total materials costs, and the escalation and sales tax allowances gives the total devices and materials costs—$15,402.00. Hardware costs have now been fully identified.

Labor Costs Labor is the next point of focus on the bid summary sheet. The number of hours estimated for rough-in of installation materials in space is 62, and the number of hours allowed for mounting and commissioning devices in space is 91. These estimates have been taken directly from the space rough-in estimate sheet. The space rough-in hours did not have to be multiplied by a factor of 1.3 before they were transferred to the bid summary sheet, since the ceilings are of the customary height. Nor did the mounting and commissioning hours have to be multiplied by a factor of 1.2. The space thermostats do not require a ladder in the process of mounting, connecting, and commissioning. The control valves at the reheat coils are preassembled (small valves can frequently be ordered preassembled), and all that is required is the connection of the pneumatic tube, most likely to a barbed connection. An allowance of 1 h each, even if a stepladder will be used, is certainly adequate. If the actuators, or topworks, of the small reheat coil control valves had to be field-installed, the 1.2 factor would be applicable. But in this case, since only a single static pressure control is involved, you can excuse it from consideration of the factor even though a stepladder is required. With experience, you will become comfortable exercising your prerogative in areas where you feel factoring is or is not appropriate.

In the equipment room, 102 labor-hours are called for, as shown on the equipment room labor and materials estimate sheet. There is no need to multiply by a factor of 1.3, for the equipment room is less than 750 ft^2. On line 12 of the bid summary sheet, 15 h is allowed for panel fabrication; this is because the panel will be bench-fabricated in the shop, thus taking advantage of the 0.8 factor permitted on the control panel estimate sheet.

In this sample job, the jobsite is 50 mi away from the shop, and so travel time must be allowed for. Exclusive of shop fabrication time for the control panel, 255 labor-hours is required at the jobsite. Allowing 8 h per day at the jobsite means that the job will require 32 workdays, which translates into 16 two-person trips to the job and back. Two workers have agreed to travel one way on their own time; that leaves 32 h to be entered on the bid summary sheet for travel time. No special requirement is envisioned on this job for checkout and supervision time, other than what is built into the standard hourly allowances for devices. Upon totaling the labor-hours, you'll find that you are looking at a job that will require 302 h. The plans include making use of a mechanic and a helper, whose hourly rates are $15.00 and $8.00, respectively. To find the cost for each, divide the required hours, 302, by 2, which is 151 h. For the mechanic, multiply 151 h by $15.00, and for the helper, multiply 151 h by $8.00. The resulting costs are $2265.00 and $1208.00, which together add up to $3473.00 total installation costs.

Travel expenses are associated with the company vehicle, which must travel 100 mi per day 16 times. (A round-trip is 100 mi.) Thus, 1600 mi at a cost to the shop of 30 cents per mile (you must determine what your actual costs are per mile) gives $480.00 in travel expenses associated with the vehicle. This includes gas, maintenance, and wear and tear (depreciation).

Engineering costs are the costs of preparing drawings and ordering materials and devices. It is reasonable to use 10 percent of total installation costs as the entry for engineering costs, since engineering time usually paces installation time at about 10 percent.

The costs for maintaining a labor force, both in and out of the office, are real, and those pertaining to a specific job should be costed into that job. Payroll taxes, insurance, and other fringes can be expressed as a percentage of labor costs. In this sample job, 16 percent is used. Labor costs are $3473.00 total installation labor costs plus $347.00 engineering labor costs, or $3820.00; and 16 percent of this figure gives $611.00 in "burden" costs. Adding up all these costs results in total labor costs of $4911.00.

Peripheral Costs The last section in the bid summary sheet focuses on peripheral costs. The size of this sample job does not warrant special consideration for cleanup crews or a construction shack or trailer. Also, no bonds are required in this case since the controls contractor is a subcontractor to the mechanical contractor (who also is a subcontractor). The electric work has been figured as part of the job, and so $150.00 is allowed for permit costs.

Mounting the actuators in the VAV mixing boxes is the responsibility of the box manufacturer, and so no factory-mounting charges apply. No subcontracts should be necessary either, since you are planning on doing both the pneumatic and electrical work required on the job. But you will have to make a submittal, for this is a plan and spec job, and so 1 percent of devices and materials costs is allowed to cover the costs of data sheets and/or brochures and necessary in-house reproduction costs.

The usual 1-year warranty is called for. For warranty costs, you can figure on 5 percent of labor and materials costs, which comes to $1015.00. This sum covers costs if you are called back during the 1-year warranty period to handle problems that may or may not be yours. All the devices you are going to order for this job are familiar to you, and so no outside technical specialist is needed. However, the owners, architects, and engineers will need a walk-through and an explanation of the system, and so time must be allotted for this. On the basis of the size of the system (and since the spec didn't state a specific number of days), you can allow 2 days at $150.00 per day for this activity. The total peripheral costs, which can now be calculated, amount to $1619.00.

You are now in a position to determine what you anticipate your raw costs will be for this project. Adding total devices and materials costs of $15,402.00 plus total labor costs of $4911.00 plus total peripheral costs of $1619.00 gives a total job raw cost of $21,932.00. Then 15 percent for overhead and 10 percent for profit are added in, to arrive at a bid price of $27,744.00. The job will go for something close to this figure.

Remember, in bidding on a job, that when your competitors need work for their shops, they will look closely at their gross margin (bid price minus raw cost) and probably adjust their bid downward. On the other hand, if they're loaded with work and the job must be done at a time they expect to be quite busy, they might adjust their price upward. You must gauge the market. Examine your shop and the need for work, consider the time frame for the work, and quote accordingly.

16

Managing the Controls Job Your Bid Has Won

IMPROVING ON YOUR BID

When you are successful in winning a controls contract, you are presented with an opportunity. Your estimate represents the plan toward profit, and of course its implementation will be part of your job management philosophy. But in the process of engineering the project, you should think of your bid as an instrument you can improve upon wherever possible while still meeting the requirements of your contract. Through profit-oriented engineering, you usually can discover possibilities for profit improvement without sacrificing quality. Management of the job includes many forms of laborsaving techniques and wise purchasing, none of which are detrimental to the thoroughness of your work, but all of which are sure to increase your profit margin by reducing job costs.

There are some specific areas related to job management and control of job costs and completion that you should consider. They will be reviewed and looked at in terms of the contribution that each might be expected to make to job profits. However, before they are discussed I would like to make some observations about presenting the work to the installers.

PRESENTING WORK TO INSTALLERS

The fact is, that for pure job rough-in, it is not really necessary for the installation crews to see the control schematics. On fast jobs, piping and wiring layouts might actually precede submittals.

As long as you have carefully prepared construction drawings showing tubing rough-in on floor-plan views, you don't need technically qualified control mechanics. All you need are workers who can run tubing. Draw your workers from the union local's labor pool, if yours is a union shop, as you need them for rough-in. Developing close ties with basic installers should not be one of your goals as a controls contractor. Release them when the work related to completed segments of rough-in is finished. For nonunion shops that may not have the security of a reserve to draw from, consider subcontracting the rough-in work (you must supply accurate construction rough-in drawings). However, you will want to develop close ties with technical experts whom you can use on a number of jobs. For copper installations, you might use the services of a plumbing subcontractor with whom you work (or wish to work) on a regular basis. The same plumbing subcontractor could also be used quite effectively for bundled polyethylene or single-line polyethylene installations. For jobs which require EMT with polyethylene, subcontract the EMT work to perhaps the jobsite electrical contractor or to an electrical contractor with whom you work regularly, and show EMT runs terminating in 4-in square boxes. This gives you the option of having the electrical contractor pull the tubes for you or having the plumber pull in the pneumatic lines. In addition, consider having the mechanical contractor set your in-line piping controls and having the sheet metal contractor set your duct-mounted pneumatic devices. By skillfully managing dependable subcontractors, you can nail down hard figures for your controls installations and then have your own salaried technicians terminate and commission the equipment.

Again, even with this approach, stay abreast of job progress and job developments by making regular visits to the jobsite to see the installation as it goes in and to answer questions that your subcontractors will surely have. Properly planned jobsite inspections won't take long, and then you're free to handle the administrative matters of your work which are vital to the profitable operation of your business. A good plan is the answer. On easy-to-read drawings, show all the installation information pertinent to the system you want. With a good plan, your subcontracted price is well founded, and your intentions for the work you want are well defined. To preserve installation dollars for strictly installation practices, remember these few dont's!

1. Don't leave installation planning to the workers in the field. You didn't estimate dollars for this.

2. Don't place installers on the job without adequate materials. Give them what they need to do the job, and pack the control devices in boxes that are adequately labeled to show the area in which they are to be installed in the building. This way the workers will know which boxes to open for the area they are working in.

3. Don't ask installers to make initial jobsite visits to determine when workers will be required. *You* do it.

4. Don't overlook early job activity, such as placing sleeves in concrete decks or underground conduit runs. Have sleeves set before concrete is poured, and watch for an opportunity to install your pneumatic lines as the mechanical contractor digs the ditch.

5. Don't place workers on the job and then neglect to supervise them. Give direction. Make suggestions. Look for shortcuts, not in quality, but in expenditures of labor. Again, know the job.
6. Don't forget to set goals for finishing certain portions of a project within a given time frame. People work better when they know what's expected of them.
7. Don't overlook the need to provide incentives for efficiency. The least you can do is to formally and publicly recognize good work. The best you can do is to include your job supervisor in the shop's profit plan. Incentive is a very important ingredient that is too frequently neglected.

The greater your efficiency as an installing contractor, the more work you will be able to win at bid time, because you can draw upon your experience as you estimate. This will allow you some latitude in reducing profit margins when bidding on projects you need in order to maintain a stable workload.

CLOSING OUT YOUR PROJECT

The final aspect of job management is concerned with selling the job to the owners, a common goal shared by you, your customer—the mechanical contractor—and the general contractor. It is in this area that many jobs go sour, because of the lack of sufficient attention to completing the lingering details required to finish the job 100 percent. A job which was won in a very competitive environment and which was well-managed up to the point of sell-off to the owner can turn out to be a problem unless you devise a plan for having the ultimate customer, the owner, accept the job and certify that you have fulfilled your contractual obligations. It's all too easy for your interest in a project to wane as the end approaches. You've been involved in the work for some time, and you grow tired of it and look forward to moving on to the next job. And yet, in the closing weeks of the work, this is the very time to intensify your efforts to take care of such things as (1) unacceptable deviations from the specifications which were pointed out by the architect or engineer, (2) control instabilities which occur as a result of poor device location or misapplication, whether the problem was created by you or not, (3) punch-list items, (4) owner's complaints about areas of the building which are either too hot or too cold, (5) in-warranty failures, and (6) proper documentation of system operation and owner orientation as required by the spec.

It is up to you to get yourself off a job by bringing it to a satisfactory conclusion and leaving behind a happy owner. And yet this aspect of your responsibilities—to yourself and to the profitability of the job—all too frequently is lacking and leads to cost overruns and dissatisfied customers. The objective is not unlike that of any business that offers a product for sale. You want to deliver a product and have it accepted and appreciated. The product you are offering is an installed system, and the plan for delivering this system to the ultimate user and obtaining written acceptance should receive top priority. Unless you plan to elicit the owner's acceptance in a timely manner, you will sustain cost overruns in labor because you were on the job too long, and the beginning of the warranty period will be delayed, causing you to be responsible to the owner for a longer period than is fair.

The plan for delivery of a good product (your system) to the ultimate customer (the owner) should begin in the early stages of your work on the project, and it should be evident in the attitude and objectives of everyone who works for you. Every person in your organization needs to realize what your objective is—delivery of a system that has been installed, within the framework of the job specifications, to the satisfaction of the customer. In a spirit of cooperation and willingness to compromise when necessary, you can overcome hurdles to your plan for delivery of this system as they arise, thereby preventing a build-up of problems to be disposed at the end of the project. Problems only intensify when they are allowed to persist. Attacking them head-on as they arise shows your determination to perform and to resolve difficulties, and creates an atmosphere of cooperation in which resentment and hostility cannot gain footholds.

Architects' Representatives

Periodically throughout the job, the architect's representatives will visit the jobsite to make sure that the work being done complies with the contract requirements. After each visit, they will usually issue a report on what they saw to all the contractors involved. If the report cites a problem with your portion of the project, don't let it anger you. Instead, right the wrong, if in fact what is written is justified, or explain why the report is in error. Occasionally, tubing specifications are not all that clear and can sometimes be read in more than one way. Where you have made an interpretation and performed in good faith, you can generally satisfy the architect's representatives with your genuineness of intent and have the reported deviation checked off. On the other hand, if you have blatantly disregarded an obvious requirement of the specifications, you must admit it and make the necessary corrections.

It is not always obvious at the time a job is being planned that the design consultant has made an unsatisfactory choice regarding the application or location of a control component. When field conditions prove that alternatives are preferable, you should be flexible in your attitude and work with the consulting engineer to relocate the control or modify the application. Personnel on your staff can make poor choices for locating devices too. For example, consider a low-limit controller for freeze protection within the casing of a rooftop unit. The shop drawings might place the sensing bulb of the low-limit controller up against the unit's heating coil. This does little toward warming the other extremities of the unit in extremely cold weather when only a little heat is necessary to satisfy the requirements of the controller, which is located right at the coil. A better location would be some distance away from the coil so that proper sampling of true casing conditions can occur. Cheerful relocation is the answer, and the only practical solution. It matters not whether the consultant or the contractor selected the original location. The point is, for the betterment of the project and for the proper protection of the rooftop equipment, the controller must be relocated. Flexibility begets flexibility. Work with others, and they in turn will be more likely to work with you.

Project Inspections by Design Consultants

Normally, as the project draws to a close, the design consultant will inspect the project and issue a punch list, which itemizes problems that were observed. The punch list should be promptly handled. When you know the allegations are accurate, correct your problems; and when you know they're not, explain what you have done and show why you would expect acceptance. I feel that the first rebuttal to a punch list should be verbal, and should be made to your immediate customer, most likely the mechanical subcontractor. The mechanical subcontractor may suggest that you discuss it with the design consultant, which you should promptly do. Hold off writing letters until you have tried to work out your differences face-to-face with the design consultant. Do not start a letter-writing duel. Discuss the issue first, and allow the design consultant merely to drop the item from further issues of the punch list. Treat the design consultant as a friend and show your desire to work amicably. Then, as word filters down through channels that your work meets with the approval of the design consultant, the matter will be dropped. When the design consultant is located many miles away, you can probably accomplish the same results over the telephone. Write a letter only as a last resort. Try to work a problem out through discussion first, and then any concessions granted by the design consultant to you need not be publicized.

Dealing with the Owner

As the owner takes over the new building, problem areas are sure to arise as people settle in and become acclimated to the new environment. Mistakes that your installers have made, such as crossing the wires to the modulating controllers or crossing the tubes to the multizone actuators, will begin to show up. When there is a problem relating to temperature control, usually the owner will contact the general contractor, who will contact the mechanical contractor, who will contact you. This will happen during the warranty period, and although you will get many calls because the owner does not understand how to operate the new system, you need to respond promptly. If a call points out a mistake in installation or application, you must make it right and be quick about it. If a call merely reflects a lack of knowledge about the system, you must educate the owner—and be quick about that too. After all, if the owner doesn't thoroughly understand the system, you didn't do a very good job of owner orientation in the first place. The better you can acquaint the owner with the temperature control system, the fewer callbacks you will have to make. And it's reasonable to assume that a certain amount of debugging will be necessary during the first few months of occupancy. The need here is to be responsive, so that the owner becomes a satisfied customer who might hire you later for additional work.

Failures during the Warranty Period

Failures during the warranty period are not too prevalent in the lines of temperature controls offered by today's major manufacturers. But they do occur sometimes. As an independent controls contractor, you will rely on your wholesale supplier to replace devices which are defective or which malfunction through no fault of your installers. Keep good records of your purchases, noting dates, order numbers, and vendor order numbers for reference if warranty claims arise. Generally, you will return the defective device either to the vendor or to the manufacturer. Some vendors may send you an in-warranty failure report form to fill out and return with the device. Others may give you instructions about returning the device using a return authorization number. Whatever the method, follow the directions and return the defective instrument promptly. Proper credit to your account usually depends on correctly returning the original control. Because it may not be practical to remove the device from the control system, you are usually allowed to return it after you have received and installed the replacement. Generally, the defect is not drastic and may consist of a deviation or sustained offset from the set point, which is more easily tolerated than having no device at all. Hence, the usual procedure is to order the replacement, with proper credit subject to the return of the original device. Keeping the customer's system in operation is the number one priority.

"When did the warranty date start on that control system we installed at XYZ Company?" Many times controls contractors have asked themselves and others this question. Your file should contain this information so that it is readily available when you need to know it. And your need to know the exact date occurs as the year progresses and problems arise close to the time when you *think* the system goes out of warranty. When you demonstrate the control system to the parties involved, it is a good business practice to have them sign a letter certifying that (1) the control system was successfully demonstrated to them on this date and (2) the temperature control system was accepted by the owners as of this date. An alternative to this, if the acceptance and demonstration procedure is fragmented because there are job delays or because not all trades are finishing at the same time, is merely to write a letter stating when the system demonstration was presented, who was there, and when the warranty period begins. This can be done in a segmented manner for large jobs, where different portions of the building are turned over to the owners at different times. Each document in your file will save you from overruns in warranty costs by establishing without question the date upon which the system passes out of warranty. From then on, the customer must expect to pay for service to the system.

As-Built Drawings

A further note on closing out a control project is in order. Most specifications and contract documents require as-built drawings to be posted on the job for future reference. These consist of your control schematics, corrected to reflect the system as it was actually installed if necessary. There can be differences between what your original schematics show and what was actually done on the job. Change orders and jobsite conditions can necessitate deviations from the first set of drawings. For example, if an owner realizes that certain needs can be better met in another manner, a change order will frequently be sent out through channels so that all trades can see what effect it will have on them and request a change order accordingly. And conditions discovered on the job to be different from what the schematics may show frequently dictate the need to modify the control connections. For example, consider VAV boxes that require a reverse-acting thermostat but that you may have thought required a direct-acting thermostat. This change should be carried over into the finalized control schematics or as-built drawings when the matter is corrected on the job. The parts numbers of the devices actually used on the job should be put on the as-built drawings as well, and the drawings should be framed and hung in the equipment room. For large jobs, those portions of the control drawings pertaining to equipment within any given equipment room should be posted there. The drawings provide a ready reference for those charged with service and maintenance.

17

Where and How to Buy Temperature Controls

Knowing where and how to buy temperature controls is every bit as important to estimating control systems as knowing how to do the estimate. In this generation of entrepreneurial controls contractors, a number of multiline controls wholesalers have sprung up, and they offer virtually all manufacturers' lines of control devices and accessories, distributing them from a central stocking source or from regionalized distribution centers. These authorized controls wholesalers may have sales offices in most sizable cities, staffed with knowledgeable personnel who conduct their trade over the telephone.

MANUFACTURERS' CATALOGS

Once you have an established account with a wholesaler, you can select complete model numbers from catalogs and then place your orders by phone or mail. If you need help in selecting parts or if you wish to verify which accessories a controller comes with, your sales adviser can usually help you complete your order by phone. The manufacturers' catalogs have been purposely designed for individual controls contractors and provide tables and charts that list complete equipment identification numbers. A good catalog will contain a part number index and a subject index. If you know the part number, you can look it up in the part number index, which will direct you to the page that offers the information you need. In the same way, you can look up the name of a control in the subject index and be referred to the page that describes it. In all likelihood the controls offered in the catalog are grouped by type, for example, pneumatic, electric, or electronic. Sufficient illustrations and wiring or piping diagrams help depict common uses of the devices offered. The better catalogs give complete technical data relating to voltage and amperage ratings, fixed or adjustable differentials, throttling ranges, maximum temperature ratings, temperature ranges, and set-point scales. For control valves, you could expect to see stated flow coefficients, mounting dimensions, closeoff ratings, types of piping connections, and temperature limitations as well as descriptions of the materials in the internal parts. A very helpful feature for those who already are familiar with one or two particular brands of controls is a part number cross-reference guide, which lists the part numbers of comparable controls put out by other manufacturers. Thus, with a good catalog and a good knowledge of the controls application business, you can do quite well at engineering and specifying model numbers for estimating or purchasing. Note that it is particularly important to keep a current catalog on hand to avoid difficulties which can arise from selecting devices that are no longer offered.

Most catalogs don't list prices for the materials you need, but price lists and discount information can be obtained from your distributor's sales assistant. Be sure of your price structure for purchasing. Correct estimating depends on this. Select a supplier that you know will handle your needs expeditiously and that will provide consistent pricing.

Most controls manufacturers offer a complete selection of instruments within the pages of their catalogs to encourage you to buy just from them. However, no one manufacturer can make everything. For this reason, each manufacturer contracts with other device manufacturers for the purchase and resale of certain controls which it is not equipped to manufacture or which it merely prefers to buy for resale and inclusion in its catalog for economic reasons. Time clocks are a good example. In any given controls catalog you may find a time clock that seems to be identical to a time clock in another manufacturer's catalog. The only apparent difference is the nameplate, and in fact it may truly be the only difference. The clock might be a product of yet a third manufacturer, and could well be available as a wholesale item under the name of the real manufacturer. Other examples of items that are often marketed under several names are interval timers, electric relays, timing switches, air dryers, filters and regulators, and air pressure gauges.

The reason for pointing out that different catalogs may carry the same item, but under a different name, is so that you can make the most economical purchases of the pieces you need. It is commonplace to find the names and model numbers of outside purchased equipment in the submittal drawings of major control companies. There have even been cases where a manufacturer's branch office has opted to purchase one or two items from an outside source because the price was lower than the in-house price. You need to be aware of price differences so that your estimating and purchasing can be every bit as effective at saving you money as are the efforts of the larger companies. Begin to develop a data sheet binder in which you file, by category, data sheets on equipment available to you at cost-efficient prices. In addition to being a very handy price reference, the binder will provide application data that will prove useful over and over again as you win bids and prepare shop drawings. The best applications people have good reference materials, and so do the best estimators.

BASIC PURCHASING CONSIDERATIONS

In selecting temperature controls for commercial applications, you need to decide how a control will be applied in order to determine what features it must contain. Much of the decision making about applications will be academic, since by the time you make your purchase you will have determined your application. You can narrow down what you need by considering the following questions, most of which apply whether the device is pneumatic, electric, or electronic.

Room Thermostats

- Will direct or reverse-acting (heating or cooling) thermostats be needed?
- What temperature range will be required?
- Should the set-point adjustment be concealed or accessible?
- Should the temperature display be visual or hidden?
- Will single-temperature or two-temperature thermostats be needed?
- Should the thermostats be locally or remotely indexed for night setback?
- What cover style and/or color will be needed?
- What voltage and amperage ratings will be required (electric systems)?
- Will special wiring (shielded) be required (electric systems)?
- Should the thermostat be snap-acting or modulating (two-position or proportional)?
- Will fan switching or system switching be required?
- Will mounting hardware be required?
- Will locking covers or thermostat guards be required?
- Should there be a fixed or an adjustable differential (or throttling range)?
- Will a transmitter or a true thermostat be required (remote setting)?

Certainly special applications can arise which require additional thought, but the list above should prove helpful when preparing application diagrams for use as submittal and installation drawings.

Temperature Controllers (Water and Air Temperature)

- Will rigid-stem or remote-bulb devices be used?
- Should they be direct or reverse-acting?
- Should the controllers be two-position or modulating?
- What temperature range is required?

- What differential or throttling range will be needed?
- What should the capillary length be?
- Should the capillary be armored or open?
- Will the devices have quick-acting or slow-response bulb fill?
- What are the sensing-bulb temperature limitations?
- Should a bulb or an averaging element be used?
- Will ambient compensation be necessary?
- Will mounting hardware (brackets, capillary clips, etc.,) be required?
- Will accessories (thermal wells or sunshields) be needed?
- Will a single or a multistage (two-position controller) be needed?
- Will remote-control point adjustment be necessary?
- Should local temperature be indicated?
- Will main air and branch line pressure gauges be needed (pneumatic systems)?
- What voltage and amperage ratings will be required (electric and electronic systems)?

Remember that in some cases, when damage to the capillary system can result in loss of thermal fill, you need to consider the position of the steam valves when the instrument goes into the fail-safe mode. Wide-open steam valves may not be desired because of uncontrollable high temperatures. On the other hand, closed valves may expose the equipment to freeze up.

Control Valves

- Will air-to-open or air-to-close valves be needed (pneumatic system)?
- Will there be local dwell or spring return in the event of loss of power (electric and electronic systems)?
- What should the flow capacity (C_v) be in the open position?
- Will straight-through or three-way valve bodies be used?
- Should the trim be bronze or stainless steel?
- Should a linear or equal percentage plug be used?
- Should renewable seats be used?
- What will the closeoff requirements (pipeline differential pressure) be?
- Should screwed, solder, weld, or flanged end connections be used?
- What will the control voltage be (electric and electronic systems)?
- Will modulating or two-position valves be needed?
- Will mixing or diverting pattern plug (three-way valves) be used?

- What will the maximum pressure rating be?
- What will the actuator spring range be (pneumatic system, sequencing)?
- Will a stem position indicator be required?
- What type of stem packing will be required?
- Will linkage or actuator mounting kits be required?

Butterfly control valves are frequently used on condenser water lines, and so special liners on the control vane may be required when tight shutoff is necessary. Generally, you don't have to worry about this, since condenser water temperature control applications don't require tight shutoff in either the tower leg or the bypass line. However, you need to be aware that seat liners are available for improved closeoff in butterfly-valve applications.

Damper Actuators

- What will the torque requirements be?
- Will direct-acting or reversing linkage be required?
- Will mounting be external or internal?
- What spring range will be required (pneumatic system)?
- Will the actuator be signal-driven or will positive positioners be required (pneumatic system)?
- What will the voltage requirements be (electric and electronic systems)?
- What accessories (crank arms, jack shafts, standoff kits) will be needed?
- Will two-position or modulating action be required?
- Will the response time be critical?
- Will high-pressure air be needed (pneumatic system)?
- Will spring-return action be required (electric system)?
- Will shaft extenders be used?
- What will the timing of the actuation period be (electric and electronic systems)?

Actuators are generally selected on the basis of the size of the damper to be controlled. Each motor is rated in terms of the total square feet of damper which it can handle. There are different ratings for pneumatic motors depending on whether the motor is to operate the damper in a two-position or in a proportional (modulating) manner. A pneumatic motor without a positioner receives a measured pneumatic output from a modulating controller, but it also can receive a sudden and distinct pressure signal from an EP or a pneumatic switching relay. A pneumatic motor that has the benefit of a suddenly increasing pressure rise to full pressure is in a much better position to overcome damper friction and resistance to movement than is a gradually positioned motor, which receives only a portion of the pressure from the modulating controller. Thus, in general, a pneumatic actuator has the capability to handle larger dampers in the two-position mode than it does in the proportional mode. This limitation can usually be overcome by using positive positioners, which have the ability to bring full main air pressure to bear upon the actuator, when required, to achieve partial movement.

Localized Control Panels

- Are they required by the spec, or are they for the installer's benefit?
- Will they be wall-mounted, unit-mounted, or freestanding?
- Will they feature system graphics, or will the fronts be plain?
- Will temperature readouts in the door be needed?
- Will a fluorescent canopy be necessary?
- Will there be enough room and enough available wall space?
- Will the piping wiring be done on the bench, or will it be done in the field?
- Will a terminal strip be advisable or necessary?
- Will NEMA-1 be okay, or will a special enclosure be required?
- Will knockouts be needed at the top, bottom, and sides?
- Will a locking door be required?

Since the introduction of the sensor-controller systems, local control panels have become commonplace on commercial pneumatic temperature control jobs. Quite often, bench fabrication and prepiping of control devices mean savings to the controls contractor, because shop labor is generally less costly and more easily controlled than field labor. The consulting engineer's drawings of the floor plans often show the location of automatic temperature control (ATC) panels, leaving little doubt that a local control panel is expected. But even when the job specs don't clearly indicate the requirement, using a local control panel in the fan room or equipment room could be to your advantage. For example, consider the need to mount EPs, PEs, and switching relays on the side of an air handler along with one or two receiver controllers. This segment of fan room work can be easily performed in your own assembly spaces, off the jobsite and away from the distractions and interferences caused by workers in other trades. Consideration should be routinely given to in-shop activities which can conceivably replace on-site labor at a savings in job cost. Project supervisors can prepare panel assembly drawings, requisition all the necessary materials, and have them ready on the workbench for the installer to assemble. The work can then be done during bad weather or when there is slack time between jobs, particularly if it is planned well ahead of the date it is needed at the jobsite. Since benchwork that is planned far enough ahead of time can be set aside if critical field work needs to be done, it doesn't interfere with the scheduling of work. You may even find that stock clerks or receiving personnel can perform not only their regular duties but also assemble panels as long as the panel assembly work by these nonunion workers does not infringe on the work under union jurisdiction done by your field installers.

PURCHASING GUIDELINES

Five families of devices should be given special consideration when you estimate and particularly when you buy: thermostats, controllers, valves, damper actuators, and control panels. Supporting devices, such as relays, time clocks, interval timers, EPs, and PEs, that are used for interlock, function switching, and interval limiting should also be selected if needed.

In purchasing control system equipment, it makes sense to buy from one manufacturer as far as is practical. It simplifies learning for both your personnel and the future owners of the systems. If you can combine instrument reliability, a good price, and the satisfactory sales engineering assistance of a local representative on a repeat basis, there is no real reason not to become an expert on one brand of equipment. Further, as you master the line, you will be able to point out the benefits of the equipment you supply if you are asked to help write a spec. Another good reason for knowing one line inside and out is that you know how long it will take between ordering and delivery. Knowing how much lead time is needed tells you when job orders should be released in order to meet installation schedules. As you develop a working relationship with one source for controls, you will become more familiar with the nomenclature used for product categories, which will make it easier for you merely to issue written purchase orders rather than spend time ordering by phone.

Stocking Control Devices

In regard to where and how to buy controls, some thought should be given to maintaining a minimum stock of commonly used items. Invariably certain control components will be needed in a hurry because (1) whoever ordered them miscounted, (2) the need for additional interlocks becomes apparent, or (3) component failure occurs, simply as a result of the law of averages, which says that when a significant number of any one device is used on a job, one or two will be defective. These commonly used devices stocked in small numbers will facilitate quick access. Some typical devices to stock are single-temperature pneumatic thermostats; switching relays; EPs and PEs; firestats; freezestats; and 24- and 120-V electric relays, perhaps in the single-pole double-throw contact configuration. Many an installer can testify to the convenience of having enough commonly used items on hand. Your stock should be well thought out and no larger than absolutely deemed necessary. Its purpose is only to save your field people some headaches when needs are urgent. Large inventories tie up capital and take up space which could be put to more productive use. But stocking adequate quantities of essential components helps reduce downtime that results when warranty-period failures and requests for service or parts can't be handled promptly. Placing some of each of the selected controls in your service vehicles will save having to make trips back to the shop.

As you order the controls for a project, keep in mind any minimum order quantities and minimum billings that your suppliers may require. There should be no problem with your major supplier. However, peripheral equipment is usually purchased on an odd-lot basis, and when you only purchase one or two items, you may have to pay a minimum charge. If this becomes a recurring problem, consider purchasing enough items to escape any less-than-minimum penalties, keeping the extras in stock, and using them, as required, on other jobs.

Delivery and Storage

Where should the goods be delivered? Your customers will usually gladly receive at the jobsite the control valves and control dampers they are to install. And if they have enough space, they may be willing to store your equipment as well. This is very important to you when specific projects limit progress payments strictly to material which is stored on the job.

When you are installing a pneumatic system, you should give some thought to having small 120-V fractional-horsepower air compressors shipped to the jobsite. These very popular home workshop candidates have been known to disappear from the job, and sometimes never to have arrived at all. Larger compressors employing three-phase electric motors are much less likely to disappear mysteriously because they are incompatible with residential electric power. Therefore, small compressors that you are going to install should be delivered to your facilities when your customer doesn't have the space to store them and the job is not far enough along for the assignment of locked storage areas.

To prevent an accumulation of job stock at your office and to eliminate the necessity of your having to maintain receiving personnel, consider renting space from a friendly contractor or associate who already has regular office hours and receiving personnel. Preferably, the person renting you the space will cage off an area for your material and, for a monthly rental fee, will receive your goods and place them in your area. Periodically, you can go by and inspect the shipments, or you can delegate this responsibility. Arranging the received material by job will make it easy for your job supervisor to pick up equipment and take it to the field for installation. With instructions given that all packing tickets be returned to the office, you can monitor the arrival of ordered goods and place vendor's invoices in line for payment. If you prefer to have the goods delivered to your place of business, and if you have the staff on hand to receive them, your office might be the better choice, but usually it is cheaper to rent unconditioned spaces for storage than to use your more expensive office space. And not having to see that someone is always on hand is an advantage, too, when your operation is small. From an economic standpoint, remote storage and receiving facilities may be most beneficial when a controls contractor is starting up business. You must look carefully at your own operation before making a decision about storage.

18

Estimating Controls Maintenance Contracts

Since the time that automatic controls were first used in commercial buildings, continuing progress in control technology has expanded the base of application. Even older systems have been refined through retrofits to provide better temperature management and greater energy efficiency, and this has increased the numbers of devices in any given building. As a result, there is a need for contractors who can provide temperature controls service for owners who do not have sufficiently qualified people on staff. Both routine maintenance and emergency service are regularly purchased by building managers, who must ensure continuing, satisfactory operation of control components. Both kinds of service are available on an as-required basis, often referred to as "spot service," and on a contractual basis under maintenance contracts. Bills for spot service are payable after the service has been rendered and include the cost of labor and materials plus overhead and profit rates. The cost of maintenance contracts is generally figured on an annual basis, payable monthly, quarterly, semiannually, or annually. For a maintenance agreement, you need to be able to estimate accurately the number of devices which the maintenance contract will cover by surveying the existing system and then forecasting how much it will cost to keep the system operational and in good condition for 1 year. From there you can quote an annual figure. In so doing, you are guaranteeing the owner a working control system at a price to which you are bound for 1 year. Many building management personnel prefer the annual maintenance contract because it allows them to forecast a fixed cost for controls maintenance. In addition, a maintenance contract assures them of service when they need it, as customers with contracts expect to receive preferential treatment over spot-service customers.

PREPARING AN ESTIMATE

Properly estimated, a maintenance agreement should provide satisfactory service for your customer and adequate remuneration for you. However, if you estimate poorly, you might quote too high and thereby lose the contract, or if you quote too low, you might receive the contract but wind up losing money on it.

In order to prepare an adequate estimate, it is very important to

1. Compile a complete list of control components in the system.
2. Know the age and condition of the system.

Carefully examine all control components in the fan room, boiler room, and equipment room. Do the same for room thermostats and above-ceiling equipment with the help of the building custodian or maintenance engineer. Inspect temperature control diagrams which may be mounted under glass or filed in the building manager's office. They will show what to look for and what not to overlook. Make a list of the devices, and their apparent or suspected condition. Look for signs of moisture or oil within the piping of a pneumatic system, usually evident around thermostats and controllers. For older systems that show signs of neglect, a preconditioning price is often quoted *before* the controls contractor will agree to accept a maintenance contract. When neglected and poorly performing systems are involved, it is reasonable to expect the owner to pay to have the system restored to the point where it can be maintained for a reasonable price. Not only will this one-time charge for system rehabilitation result in a lower and more reasonable maintenance contract cost, but also the contractor will be more likely to accept contract renewals and the building owner can expect consistent annual renewal costs.

TYPES OF MAINTENANCE CONTRACTS

There are two types of maintenance contracts. One offers preventive maintenance, and the other, which is more comprehensive, offers general maintenance. The preventive maintenance contract specifies the number of visits the servicing technician will make, enumerates the devices to be serviced, and identifies the preventive services to be performed. Any replacement equipment is billed as an extra, as well as any emergency service that may be needed. The general maintenance contract, on the other hand, not only specifies the number of visits and services, but also covers the cost of replacement equipment and emergency service. Sometimes a distinction is made between straight-time emergency labor and overtime emergency labor, with the stipulation that in the case of overtime labor, the owner will pay the difference between the standard and overtime rates. In such cases, the agreement usually states what the additional cost per hour will be. Since both the preventive agreement and the general maintenance agreement are contracts, your attorney should be involved in their preparation. However, even if your attorney will be involved, it is a good idea at this point to examine the mechanics of each kind of agreement to see how they differ and to identify certain features that should be included.

After looking at the following examples of preventive and general maintenance agreements, you will have a better feel for what your building survey should cover in regard to preparing the estimate.

The Preventive Maintenance Agreement

Form 18-1 is a sample preventive maintenance agreement.

PREVENTIVE MAINTENANCE AGREEMENT

ABC Controls Company, hereinafter referred to as the contractor, offers certain temperature control preventive maintenance service, which is specifically described in the following terms and conditions of this instrument for the XYZ Building at ___ , ___ , ___ , in connection with only those control devices itemized in Schedule A hereto, for the net annual fee of $_____ , payable monthly.

Under this agreement, the contractor shall inspect, clean, adjust, and calibrate as necessary the controls and control components as listed in the hereinbefore-mentioned Schedule A twice during the calendar year, once at the beginning of the cooling season and once at the beginning of the heating season. In connection with service to these instruments, the contractor shall be free to start and stop equipment as may be required to ascertain satisfactory operation of the controls.

Should the replacement of control components become necessary, as determined by the contractor, for the proper performance of the system, such replacement shall be at additional cost to the acceptor of this agreement, and shall be billed at the contractor's cost plus 25 percent, and additional labor for such replacement work at the hourly service rate of $_____ , except that the acceptor shall first be notified of such need for replacement and shall further have the right of refusal, such replacement work being at all times subject to the approval of the acceptor. Additional work thus approved shall be evidenced by extra work orders in each case, written and signed by the acceptor.

Preventive maintenance and any extra work as may be authorized shall be performed during regular working hours for the contractor's personnel, and any need for after-hours, overtime, or weekend work that may arise and be approved by the acceptor shall be billed as an extra to this agreement at the hourly service rate of $_____ , multiplied by 1.5 for work that is done after regular working hours Monday through Friday, exclusive of holidays, and not exceeding 6 hours extra for any 1 day. Work done on holidays, Saturdays, or Sundays and work in excess of the first 6 hours

Form 18-1 Preventive maintenance agreement. This form continues on pages 186 and 187.

of extra work after regular working hours Monday through Friday shall be billed at the hourly service rate of $_____ , multiplied by 2.

Labor expended in the performance of the preventive maintenance for the control system shall be itemized by the service technician through the use of a service report and shall be signed by the acceptor's representative to certify services performed. A copy of this report shall be left with the building management personnel. Each report shall clearly indicate the services performed that day and shall enable the building management personnel to track services rendered.

The period covered by this agreement is 1 year from the date of acceptance appearing below, and renewals shall be through the execution of a new agreement only.

Schedule A (List of controls covered by this agreement)

Air Handling Unit 1

Two (2) firestats, one (1) freezestat, one (1) mixed air low-limit controller, one (1) discharge temperature sensor, one (1) 3/4-in three-way hot water heating coil control valve, one (1) 2-in three-way chilled water cooling coil control valve, three (3) damper actuators, and three (3) control dampers, one in the fresh air, one in the return air, and one in the relief air. Also, in the local control cabinet, one (1) receiver controller, one (1) 120-V ac EP, one (1) switching relay, and one (1) minimum position selector switch for damper control.

Hot Water Converter

One (1) receiver controller, one (1) hot water temperature transmitter, one (1) outside air temperature transmitter, one (1) EP, one (1) 1 1/4-in steam valve, one (1) hot water temperature receiver gauge, and one (1) PE.

Water Chiller

One (1) outside air changeover thermostat, one (1) chilled water and one (1) condenser water flow switch, one (1) chilled water temperature sensor, and one (1) receiver controller for operation of the pneumatic actuator for capacity control. (Neither the chiller itself nor its controls furnished by the unit manufacturer are covered.)

Condenser Water Cooling Tower

One (1) condenser water temperature sensor, one (1) receiver controller, two (2) PEs and one (1) 6-in three-way control valve.

Space Temperature Controls

Twenty (20) wall-mounted thermostats and twenty (20) 3/4-in three-way reheat coil control valves.

Compressed Air

One (1) 1-hp 208-V three-phase single-motor air compressor with 30-gal storage tank, inclusive of starter and pressure switch, belt, and pressure relief valve; one (1) 1/5-hp 120-V 10-cfm moisture condenser; and one (1) combination filter and pressure-reducing station.

Devices and equipment not specifically listed in the foregoing Schedule A are not covered by this agreement.

This preventive maintenance agreement is a temperature controls maintenance contract and does not cover mechanical equipment, such as boilers, converters, pumps, and cooling tower.

In testimony of their agreement as hereinbefore given, the parties hereto have hereinafter affixed their signatures this the ____ day of __________ , 19__ .

ACCEPTOR:	OFFEROR:
Company ____________________	Company ____________________
By ____________________	By ____________________
Title ____________________	Title ____________________

As mentioned before, for any contract you plan to enter into, retain competent legal counsel. Form 18-1 is intended to offer guidance as to scope, but your attorney should approve the form which you develop for regular use.

If the itemization of devices in Schedule A seems to border on overkill, remember that there are many aspects to the mechanical systems in today's commercial buildings, and Schedule A protects you from having to maintain equipment you didn't plan to include. Also, control modifications or additions made by building maintenance personnel could become your responsibility without such a list.

The General Maintenance Agreement

A general maintenance agreement is similar to a preventive maintenance agreement but more inclusive in terms of component replacement and the rendering of emergency service. The contractor assumes a greater degree of risk, and the building managers and owners are freed from worrying about controls. Consequently, the customer expects to pay more *and* receive more in the way of services. Form 18-2, which follows, is a sample general maintenance agreement.

GENERAL MAINTENANCE AGREEMENT

ABC Controls Company, hereinafter referred to as the contractor, offers certain temperature controls general maintenance service, which is specifically described in the following terms and conditions of this instrument, for the XYZ Building at ___ , ___ , ___ , in connection with only those control devices itemized in Schedule A hereto, for the net annual fee of $_____ , payable quarterly.

Under this agreement, the contractor shall inspect, clean, adjust, and calibrate as necessary the controls and control components as listed in the hereinbefore-mentioned Schedule A twice during the calendar year, once at the beginning of the cooling season and once at the beginning of the heating season. In connection with service to these instruments, the contractor shall be free to start and stop equipment as may be required to ascertain satisfactory operation of the controls.

Additionally, should the replacement of control components become necessary, as determined by the contractor, for the proper performance of the system, such replacement shall be at no additional cost to the acceptor of this agreement, except that the contractor shall always have the option to repair existing devices when it is more economical to do so. Such replacements or repairs shall be made as quickly as parts can be obtained.

Work under this agreement shall be performed during regular working hours for the contractor's personnel, except that emergency repairs shall be performed as quickly as the contractor can respond with personnel and material, regardless of the time of day or the day of the week. The acceptor agrees to limit the call for overtime work to true emergency situations, when the proper maintenance of temperature controlled conditions is vital to the acceptor's personnel or equipment.

Labor and materials expended in the performance of this general maintenance agreement for the control system shall be itemized by the service technician through the use of a service report and shall be signed by the acceptor's representative to certify services performed. A copy of this report shall be left with the building management personnel. Each report shall clearly indicate the services performed that day and shall enable the building management personnel to track the services rendered.

The period covered by this agreement is 1 year from the date of acceptance appearing below, and renewals shall be through the execution of a new agreement only.

Schedule A (List of controls covered by this agreement)

[The various control components would be itemized here as in the preventive agreement.]

Form 18-2 General maintenance agreement. This form continues on page 189.

Devices and equipment not specifically listed in the foregoing Schedule A are not covered by this agreement.

This general maintenance agreement is a temperature control maintenance contract and does not cover mechanical equipment, such as boilers, converters, pumps, and cooling tower.

In testimony of their agreement as hereinbefore given, the parties hereto have hereinafter affixed their signatures, this the ____ day of __________, 19__ .

ACCEPTOR:	OFFEROR:
Company ____________________	Company ____________________
By ____________________	By ____________________
Title ____________________	Title ____________________

Many companies today offer total building maintenance contracts, which include maintaining equipment as well as controls, but in some cases, the controls expertise is less than what the owners could obtain by purchasing just a temperature controls maintenance contract. For this reason, as a specialist in the field of controls, you should offer a specialized maintenance agreement so that the owner can utilize the services of experts to ensure quality service for the controls portion of the system.

Now that you have examined the agreements under which you would perform, you can study the procedure for estimating an acceptable maintenance contract price.

PERFORMING PREVENTIVE MAINTENANCE

You must familiarize yourself with what preventive maintenance involves. Then you can better understand the structure of the estimate sheet, which will give you an idea of how much time to allow for preventive maintenance functions. Under both the preventive and the general maintenance contracts, preventive measures are used and form the best safeguard against your having to handle a crisis situation in the customer's temperature control system. The customer in a maintenance contract deserves the maintenance work that has been promised. And a properly maintained system generally remains a working system, free from surprise difficulties that create crisis scheduling for your service personnel. And so not only do you owe it to your customers to perform proper maintenance, but you owe it to yourself—you stand to profit when you take care of the system because you will be preventing major problems.

As an example, assume you discovered, during a routine maintenance check-out, that the minimum fresh air damper on an air handler was stuck open because the linkage was broken. Needless to say, you would repair the linkage. However, if this problem remained undetected, an extremely cold night in the winter would likely produce a frozen coil, and the owner would surely wonder why the coil froze. If this was the case, who would pay to have it fixed? At the very least, it would be embarrassing for you since you are supposed to be maintaining the system. Of course, the linkage could break anytime, even right after you make a routine check. But the point is that properly performed maintenance work will catch items that can save you money and embarrassment and can prevent headaches for your customer.

Now let's look at the controls found in a building and review some of the things to do.

Room Thermostats

Each thermostat should be "ear-checked" for calibration and operation. For pneumatic thermostats, this kind of checking can consist of slowly adjusting the set point from one extreme to the other and listening for the bleed or build of the branch line pressure. It should occur close to room temperature. This method is quick, and when the make-break point is close to the actual room temperature, further calibration should not be necessary.

Consider a single-temperature pneumatic thermostat with a temperature range of 55 to 85°F. If the actual room temperature is 70°F, then as the set-point indicator is moved past 70, the output from the thermostat will bleed or build, depending on its action, and you can hear this. This tells you that the thermostat is responding to adjustment of the set point as the indicator crosses actual ambient conditions. Make sure to listen closely to see that the bleed or build of pressure is not sudden and short. There should be some capacitance in the valve, damper actuator, or PE that the thermostat controls, owing to the reservoir within the actual controlled devices that must be filled and exhausted. But if the branch line flexible lead from the thermostat has become crimped, access to the controlled devices reservoir is blocked, and the amount of control air being bled or fed can be a telltale sign. Thus, ear checking often uncovers problems that must be checked further.

Summer-winter thermostats need not be indexed from one mode to the other because you will check them out twice a year anyway, once when the thermostat is in the heating mode and once when the thermostat is in the cooling mode. Day-night thermostats, however, should be switched over, and their nighttime calibration point should be checked as well.

Electric thermostats also can be checked by adjusting the set point if they are snap-acting; in this case, you can often ear-check them by listening for the closure of the contacts, or you can remove the cover and watch the circuit opening and closing in response to your adjusting the set point past room ambient conditions. Proportional electric thermostats can be observed with the cover off for midpoint positioning of the slide-wire as you move the set point to actual room temperature. Electronic temperature sensors mounted on the wall can be checked at the controller by monitoring the controller's output as you index the controller set point to the actual room temperature.

Unit-Mounted Thermostats

On unit ventilators and fan coil units, you can most likely see, hear, and/or feel discharge temperature changes as you move the adjustment dial in the vicinity of the returning air temperature, as indicated by your test thermometer. Both pneumatic and electric thermostats are probably easier to check when they are unit-mounted because you can witness the response of the devices within the unit. However, many of the unit-mounted thermostats do not display the set point in degrees. Instead, the dials are frequently marked "cooler-warmer." Returning air temperatures reaching the unit at floor level shouldn't be shown in degrees on the face of a unit-mounted control, because occupants are misled into thinking that the set point means the temperature they feel, which is higher than the temperature at floor level. So a dial marked "cooler-warmer" or "lower-higher" presents a better alternative to indicating the set point in degrees on return air controlling devices.

Using a test thermometer, take a reading of the air temperature at the location of the sensing element for the return air thermostat. Then rotate the knob or dial to observe the thermostat's response, preferably in its midrange. This will be a good indication of whether the thermostat is in control, particularly if the room occupant has been able to control the temperature satisfactorily. If the room temperature is unsatisfactory, insert a branch line pressure gauge for pneumatic thermostats, or use a meter for electric or electronic controls, and see if the thermostat is really calling for heat if the room is too cool, or cooling if the room is too warm. You should do further sequence checking to determine why the unit has not responded to change the temperature in the room. In such a case, your preliminary check has led you into needed troubleshooting.

Sensors and Receiver Controllers

Duct-mounted, pipeline-mounted, and troffer-mounted temperature sensors should be checked for proper temperature indication. For pneumatic transmitters, determine the temperature range of the sensor, and check its output by reading the signal pressure reaching the controller and comparing that pressure reading with a temperature-pressure chart that shows what the output should be for any given temperature. A linear pneumatic transmitter with a 3- to 15-lb/in^2 output over a given range can be expected to produce a computed signal in line with the pressure shown in Table 18-1. This is the first step in proving the accuracy of a transmitter, for until you know that its output is correct, you should not attempt to correct the reading at a temperature gauge on a local panel or to calibrate the controller receiving the signal. Four temperature ranges in degrees Fahrenheit are offered in Table 18-1, and you can see at a glance what the pneumatic pressure should be for any given temperature in any of the ranges. With transmission indication systems, the greatest accuracy occurs at midrange, and for that reason various ranges are offered by manufacturers

Table 18-1 • Pneumatic Temperature-to-Pressure Conversion Chart

Temperature Range of Sensor, °F				
0 – 100	0 – 200	50 – 150	50 – 250	Output, lb/in^2
0	0	50	50	3.0
4	8	54	58	3.5
8	16	58	66	4.0
12	24	62	74	4.5
16	32	66	82	5.0
20	40	72	90	5.5
24	48	76	98	6.0
28	56	80	106	6.5
32	64	84	114	7.0
36	72	88	122	7.5
42	84	92	130	8.0
46	92	96	138	8.5
50	100	100	146	9.0
54	108	104	154	9.5
58	116	108	162	10.0
62	124	112	170	10.5
66	132	116	178	11.0
70	140	120	186	11.5
74	148	124	194	12.0
78	156	128	202	12.5
82	164	132	210	13.0
86	172	136	218	13.5
90	180	140	226	14.0
94	188	144	234	14.5
98	196	148	242	15.0

to permit selection of a range that includes what the normal control point will be. The sought-after set point will be generally in the middle of the range.

Many manufacturers produce pressure gauges that offer good accuracy and feature multiple temperature ranges on their face. Then, both the pressure and the temperature that relates to it can be read directly on the gauge. Personally, I prefer to use a two-part system: one good, very accurate test gauge with at least a 4-in face that reads in tenths of a pound and a separate temperature-to-pressure chart that lists the appropriate pressure for each degree in each temperature range. I find them easier to use than multirange pressure gauges.

For electronic temperature transmission, use a good digital volt-ohm meter that also permits you to read milliamperes of current flow. You can take readings of the sensor input to the controller and find the temperature from a predetermined scale. Then, you can compare the temperature you found using a good test thermometer with the temperature indicated by the sensor transmitter.

Once you have determined that the temperature input signal is valid, rotate the controller adjustment set-point knob; a reaction should occur as you swing past the input temperature being received at the controller. These quick checks are not substitutes for thorough calibration procedures, but they do allow you to validate sensor stability and will usually save you from having to do a more complete checkout.

Pressure controls, including static pressure devices, should be tuned to actual conditions for observation of reaction. Where moving the set point will upset control conditions, obtain permission for temporary shutdown while you conduct your tests. Static pressure controls can be pumped *gently* with a squeeze bulb, and a static pressure gauge (reading in inches of water), or a manometer, can be teed into the line to verify controller accuracy. Receiver controls can be pumped with a squeeze bulb, too, to test their ability to bleed and feed their branch lines at a point that corresponds with the set point.

Remote-bulb thermostats, either electric or pneumatic, can easily be checked by rotating the set-point indicator. Take care that such action does not start up or stop equipment whose operation must not be disturbed. For an electric controller, remove a wire or add a jumper to prevent unwanted starting or stopping of equipment. For a pneumatic control, remove a line that must not be pressurized, or use a substitute means for maintaining the signal in a line that must be pressurized.

Relays and Switches

Actuate pneumatic diverting relays by manual signal indexing and then observe their response. Watch for pneumatic switching relays that hang up or do not properly isolate exhausted lines. Listen for continuing bleed, which tells you that internal ports are not sealing off. Manually operate diverting and gradual positioning switches to make sure that their detents position the switch with good alignment, allowing tight sealoff of blocked lines and atmospheric dumping of exhausted lines. Tumblers and cams of electric switches with two-positioned knobs or handles should be tight, giving a clean action with no knob wobble. EPs should energize properly with little or no hum, and PEs should hold pressure without the need for continuing feed from their controller to maintain contact engagement.

Damper Actuators

Put the damper actuators through their stroke. Check to make sure that all the dampers open and close properly and that the linkages are free to work smoothly without binding or halting action due to lack of lubrication. Check for run-on in electric actuators, which could be a sign of stripped gears. In pneumatic actuators, listen for leaking of pneumatic pressure when the motor is fully pressurized. Swivel-mounted actuators should be free to pivot so as to maintain a position of maximum force with respect to the damper crank arm. Dampers should seal tightly, and blades should turn freely and rotate evenly. Check positive-positioning devices on damper motors to be sure their feedback mechanisms are workable. If they're not, the dampers under their control may be two-positioning instead of modulating. The full 90° arc should be exhibited as the damper crank arm is stroked from one extreme to the other. Are linkages disconnected? If so, this could be a sign of a nonworking control that building personnel manually overrode.

Control Valves

Close the valve and check to see if the heating coils under its control cool down. Check the temperature of the air leaving the coil with the valve in the closed position, and compare it with the temperature of the air entering the coil to make sure that leak-through is not occurring. On chilled water coils, watch for a rise in the temperature of the air leaving the coil when the control valve is positioned to stop flow through the coil. Apply valve stem lubricant periodically, and be particularly alert to possible pneumatic diaphragm failures on valves used in high-temperature applications. When steam or water is leaking around the valve stem, replace the packing, using the manufacturer's recommended valve repacking kit. On old electric valves, look closely at the mechanical linkage which adapts rotary output from the actuator to up-and-down stem movement for the valve. Quite often, the stem only has to travel an inch or less, and a worn linkage may not be able to fully open or close the valve. Where the linkage is dry, apply lubricant as recommended; then make sure that the action of the valve is still correct. Be sure that valved-off control valves haven't been affected by failure at the controller, which could cause the owner to assume that the only way to control the system is to manually close the isolation valves that are intended for use only when repairs are needed. In all cases, for both electric and pneumatic control valves, look for *heavy* corrosion which might signal body failure.

Air Compressors

The air compressor—the heart of the pneumatic system—must be maintained in good working condition, and this includes the dryer, the filter, and the pressure-reducing station. Look for evidence of oil in the air lines, visible in the polycarbonate reservoirs of filters and traps. Additionally, make sure that water is not accumulating in traps that should be dumping it. If the trap in the air dryer contains water, then the dryer is not operating at a condensing temperature that is low enough to remove all the water, and/or the trap is not "springing" as it should or is stopped up with sludge. Some air dryers have openings into which you can insert a test thermometer with a dial face and a long, slender metal probe. Read the thermometer to see what temperature the high-pressure air in the dryer is being reduced to. A temperature between 35 and 38°F is necessary to purge the air of moisture and yet prevent freeze-up in the exchanger. If possible, valve off the air compressor and then dump enough air from the tank to cause it to come on. Check to see if it recovers quickly or if it runs unnecessarily long just to replace the compressed air exhausted from its reservoir. Of course, you should also stop the unit and check the oil level. Check belt tension, and be sure that the motor is receiving proper lubrication. If there is oil carryover, you may need to replace the rings.

The preventive maintenance checks just discussed are the minimum of what you must do in terms of requirements, but they should give you an idea of what needs to be done during routine preventive maintenance.

ESTIMATING CONTROLS CONTRACTS

The two forms discussed in this section provide a logical way to estimate the cost of both preventive and general maintenance agreements. Form 18-3 is a sample preventive maintenance survey and estimate, and Form 18-4 is a sample general maintenance estimate. Form 18-3 is presented first and deals with the equipment you will be responsible for maintaining and the number of labor-hours you can expect to spend on the job during a 1-year period. This preventive maintenance survey and estimate is a stand-alone estimate for preventive maintenance contracts, and it is the first part of the estimate for general maintenance contracts. So no matter which type of contract you are surveying for, Form 18-3 is a very important ingredient, because in both types of proposals, you are going to provide the maintenance for the control system. Thus, in each case, you need to know how many preventive maintenance hours are likely to be required.

SURVEYING A SYSTEM

The initial survey of a control system must be thorough enough and sufficiently documented to permit correct pricing and proper labor planning. Here Form 18-3 is useful because it is designed to help you identify all the devices in the system you are surveying. In doing your survey, always ask to see the control diagrams that go with the system. They are probably posted on the wall somewhere in the equipment area. Check them against the control components that are being used on the job now, and make sure that you list all the control components on Forms 18-3 and 18-4. Also, be sure to answer the reconditioning survey questions. These can help you decide whether the system should be partially overhauled before you agree to maintain it for a year. (This is particularly important in regard to general maintenance contracts.)

The first part of Form 18-3 is a survey, and it prompts you to consider the existing system in terms of its serviceability and its accessibility to your

PREVENTIVE MAINTENANCE SURVEY AND ESTIMATE

Job name ____________ Address ____________
Date of survey ______ Dwellings on job? ______ Yes ______ No ______
Survey by ____________ Owner's rep ____________
Pneumatic or electric system? ____________ Age ____________
Essentially as-built? ____________ Prior maintenance? ____________
Distance from office ____________ Central monitoring? ____________
Oil or water in lines (pneumatic systems)? ________ Extent ________
Damper linkages loose? ____________ Extent ____________
Readouts correct? Pneumatic ______ Electric ______ Both ______
Occupants comfortable now? ________ Local or general? ________
System used as designed? ________ Intentional defeats? ________
Congested or outlying area? ________ Tolls and parking ________
Excessive compressor run time? ________ Cycles per hour ________
Good occupancy record (leased spaces)? ________ Why? ________

Controls Takeoff Summary		**Semiannual hours**	
Quantity		**Each**	**Total**
________	Room thermostats, pneumatic	0.2	________
________	Room Thermostats, electric	0.15	________
________	Unit thermostats, pneumatic and electric	0.3	________
________	Mixing boxes, control valves, or VAV boxes	0.3	________
________	Reheat coils, electric or wet heat	0.4	________
________	Temperature transmitters, electric and pneumatic	0.5	________
________	Receiver controllers, electronic and pneumatic	0.75	________
________	EPs, PEs, and other electric and pneumatic relays	0.2	________
________	Switches, electric and pneumatic	0.15	________
________	Damper actuators and linkages	1.0	________
________	Control valves, 2 in and under	0.5	________
________	Control valves, 2 1/2 in and over	0.75	________
________	Pressure and receiver gauges (meters)	0.3	________

Form 18-3 Controls preventive maintenance estimate. This form continues on page 197.

	Item	Hours	
________	Dampers, 16 ft^2 and under	0.8	________
________	Dampers, over 16 ft^2	1.2	________
________	Air compressors, 3/4 hp and under	1.5	________
________	Air compressors, over 3/4 hp	2.0	________
________	Air dryers, 10 cfm and under	0.3	________
________	Air dryers, over 10 cfm	0.5	________
________	Electric remote-bulb controllers	0.2	________
________	Pneumatic remote-bulb controllers	0.3	________
________	Enthalpy controllers, morning start-up optimizers, and static pressure controllers	2.0	________
________	Firestat or freezestat	0.2	________
________	Duct smoke detector	0.5	________
________	Flow switch, differential or positive displacement	0.25	________
________	Time clocks	0.15	________

Total hours		________
Multiplied by 2 for annual hours (______ x 2) =		________
Annual hours ______ x hourly rate ______	= $__________	
Annual hours ______ x hourly rate ______	= $__________	
Annual travel ______ x hourly rate ______	= $__________	
Annual travel ______ x hourly rate ______	= $__________	
Total direct labor costs		$__________
Burden at ______% of direct labor costs		+ $__________
Annual mileage at 30 cents per mile		+ $__________
Subtotal: Anticipated annual contract costs		$__________
________% overhead		+ $__________
Costs plus overhead		__________
________% profit		+ __________
Costs plus overhead and profit (bid price)		= $__________
Shading factor (if applicable)		x__________
Total preventive maintenance contract bid		= $__________

service personnel. For example, it is important to know if the system that was originally installed has been modified. If the system has not been changed, then the original schematic drawings of the system that you hope to find on the job should provide you with a schematic picture of the system. If the system has been modified, then you will be faced with some lost time as your workers try to discover how the system was changed. Knowing if the system has had prior maintenance has bearing upon the estimate since you can expect that a system that has been kept up will be less costly to maintain than one that has not been maintained. Knowing how far the job is from your office is also important, for travel time must be figured into the probable costs.

Another thing you'll want to know is whether the system is being monitored at a centralized computer-based control center, which permits remote survey of the points in the system. If so, two things come to mind. First, the console operators can aid your service personnel, as they ferret out problems in the job, by spot-checking temperatures, and their knowledge of the points in the system can help familiarize your personnel with the project. Secondly, if the contract is just a temperature controls maintenance proposal, you need to be sure, for the sake of clarity, to exclude the central automation system in the control center from maintenance. If your shop is equipped to handle central automation systems, you could elect to include the automation system as well. But here we are talking about temperature controls only.

The presence of oil or water in the air lines signals problems at the compressed air source (for pneumatic systems). Select a thermostat on a lower level of the building fairly close to the air compressor. Ask permission to remove it, or have the owner's representative remove it, so that you can inspect the airhead behind it for the presence of oil or water. Either will usually find its way to devices located on the lowest levels of the building first, and so this is where you want to check. If water is present, plan on purging the lines with clean, dry air, and suspect that the air dryer is not doing its job. If oil is present, plan on evacuating the lines with refrigerant gas and then flushing with clean, dry air; and check that compressor!

Often when ventilation cycle controls go on the blink and are not repaired, the building personnel will disconnect the damper linkages to prevent the introduction of cold air in the winter and hot, humid air in the summer. If you find that the linkages are disconnected, you can suspect that some devices need to be replaced or calibration has to be performed.

Next, look at the readouts of the temperature gauges (for pneumatic systems) or the temperature meters (for electric systems). Study them briefly to satisfy yourself that what you are seeing represents a reasonable display of temperatures. If some of the displays are registering zero, disconnected lines could be the culprit. If some are registering full pressure, blocked leak ports in pneumatic sensors may need to be attended to. When there is local and/or central display of remote temperatures, count the number of displays and enter the appropriate count in the "Pneumatic" and "Electric" spaces. This will help you confirm your count of sensors and transmitters later on. If both the pneumatic and electric readouts need attention, check the "Both" space to remind you that your technicians need to be versed in both pneumatic and electrical transmission technology.

Ask questions about temperatures in the building. Are the occupants of the building comfortable? Is poor temperature control a problem in just certain areas, or is it a widespread situation? If the problem is confined to certain areas, the trouble is probably in the local terminal units, while if the problem is widespread, the trouble is probably in the central equipment. Ask questions about the system itself. Is it being used as it was designed? Are the manual override features in the system being taken advantage of? You may find that reeducating the owner's operating personnel on how to operate the system can save you money during maintenance. Oftentimes a person who doesn't understand how to operate a device, and who is reluctant to ask for help, will eliminate the device as a stumbling block by intentionally defeating its operation. Controls that have been jumpered or removed and wedges that have been inserted under pivotal levers of controllers are signs that portions of the system are being intentionally defeated.

Also consider the area where the building is. If it's in a congested area, you need to allow time for your technicians to park their vehicles and tote their equipment to the building in the morning and then to pack up their equipment and get back to their vehicles at the end of the day. And tolls and parking costs must be allowed for too.

Finally, make it a point to look at the compressor on your initial survey visit. Stand there long enough to witness a few on and off cycles. Note how may times it started in the course of 30 min, how long it ran, and what percentage of time it was on in those 30 min. This will give you an idea of the compressor's capability of serving the presently connected pneumatic controls. Your observations might suggest leaks in the system if no additions have been made and the compressor was large enough for the original system. Or you might find that the unit needs new rings, or that not only is it pumping air too slowly, but it is also pumping oil vapors into the lines. Finally, if the building is a leased-spaces building, note, as tactfully as you can, whether all the spaces have been leased, and try to determine, perhaps later, if the occupancy rate has been good in the past. If the occupancy rate is poor, ask yourself why. Better yet, if possible, ask someone who was an occupant in the building.

The sole purpose of the first part of Form 18-3 is to help you decide whether you want to maintain this system under contract. The items and areas listed for you to consider are meant to help you direct your thinking about the system. It is useful to rate the desirability of maintaining the system on a scale of 0.7 to 1.6, which is a spread of 10 percentage points. The number, within this range, that you give to the system under consideration becomes your "shading factor." This factor is the last arithmetical function on Form 18-3. If you use 1 as the shading factor, the hours per unit as listed become your estimate. If you use 1.2, you intentionally raise the time anticipated, on the basis of the time you feel you'll need to maintain the system. If you use 0.8, you lower the anticipated time you feel you'll need.

The next step is to sketch the control loops, the same as you would do for estimating new work. If control drawings are available, you're way ahead of the game. Once you have the sketches and know how the system works, enter the quantities beside the appropriate control devices, dampers, and terminal units on Form 18-3. If a control unit is not specifically listed, log it in, with the family it belongs to or list it on a separate sheet of paper. Multiply the quantity of each device by the number of hours estimated to service it semiannually. Add these figures and multiply this total by 2 to find the number of hours estimated to service it annually.

Two different lines are allowed for determining the hourly rate; one lets you take into account the use of journeymen with helpers, and the other lets you figure the cost if the rate increases during the period of the contract. Once you know your total direct labor costs, enter your personal burden factor (which was assumed to be 16 percent in the estimates for new work) and multiply it by the total direct labor costs. Burden covers such items as insurance, payroll taxes, unemployment, pension, and vacation fund. If the job is some distance away, figure how much mileage you can expect to incur in a year's time, and multiply that figure by 30 cents per mile (or whatever cost per mile you know to be more accurate). Add to find your anticipated annual contract costs. Enter your overhead and profit, and finally figure in the shading factor, so that this estimate will be based on your observations and takeoff.

One word of caution! In a system which is obviously in trouble for lack of maintenance, hold off making your estimate until you have been permitted to recondition the system on a time and materials basis. Do a good job of reconditioning and encourage your service personnel to look for problems and report built-in design flaws. You'll have a better feel for the system, and you can be up front with your customers about the inability of even good maintenance to overcome an inherent design difficulty. As a result, you'll win the contract, you'll already be familiar with the system, and your customer will know what to expect.

ESTIMATING A GENERAL MAINTENANCE AGREEMENT

Under a general maintenance agreement, you assume the responsibility for replacing control components and making emergency service and unscheduled service calls. Therefore, from an estimating standpoint, the basic difference between a preventive maintenance agreement and a general maintenance agreement is that the general maintenance agreement covers the costs of replacing control devices and of providing additional labor for emergency and unscheduled service. As mentioned earlier, the preventive maintenance survey and estimate (Form 18-3) is the first part of the general maintenance estimate (Form 18-4), and is used to find the total anticipated costs of general maintenance. Form 18-4 is presented next and is used to forecast unscheduled costs.

GENERAL MAINTENANCE ESTIMATE

Cost of Controls

Quantity	**Cost, Each**	**Total Cost**
__Pneumatic wall-mounted thermostats	\$____	=\$________
__Electric wall-mounted thermostats	\$____	=\$________

Form 18-4 Controls general maintenance estimate. This form continues on pages 201 and 202.

___Pneumatic unit-mounted thermostats	\$_____	=\$__________
___Electric unit-mounted thermostats	\$_____	=\$__________
___Terminal unit controls	\$_____	=\$__________
___Reheat coil controls	\$_____	=\$__________
___Pneumatic temperature sensors	\$_____	=\$__________
___Electronic temperature sensors	\$_____	=\$__________
___Pneumatic receiver controllers	\$_____	=\$__________
___Electronic receiver controllers	\$_____	=\$__________
___Pneumatic and electric relays	\$_____	=\$__________
___Switches, pneumatic and electric	\$_____	=\$__________
___Pneumatic damper motor and linkage	\$_____	=\$__________
___Electric damper motor and linkage	\$_____	=\$__________
___Control valves, 2 in and under	\$_____	=\$__________
___Control valves, 2 1/2 to 4 in	\$_____	=\$__________
___Control valves, over 4 in	\$_____	=\$__________
___Gauges (all)	\$_____	=\$__________
___Compressor pump and motor	\$_____	=\$__________
___Air dryer, pressure relief valve, and filter set	\$_____	=\$__________
___Pneumatic remote-bulb thermostats	\$_____	=\$__________
___Electric remote-bulb thermostats	\$_____	=\$__________
___Enthalpy controllers and morning start-up optimizers	\$_____	=\$__________
___Static pressure devices	\$_____	=\$__________
___Firestats and freezestats	\$_____	=\$__________
___Duct smoke detectors	\$_____	=\$__________
___Air and water flow switches	\$_____	=\$__________
___Time clocks and override timer	\$_____	=\$__________
___Industrial controllers	\$_____	=\$__________
______________________________	\$_____	=\$__________
______________________________	\$_____	=\$__________
Total current estimated value of system's control components		=\$__________

Do not include thermostat rough-in fittings, air tanks, dampers, panel enclosures, or tubing and wire.

Controls Annual Replacement Costs

Total current estimated value of system's control components $____

Components cost factor selected from table below ×____

=$____

_% overhead +$____

=$____

_% profit +$____

=$____

Projected annual cost of replacing components $____

Components Cost Factors

Age of System, yr	Prior Maintenance: Good	Poor	None	If First Recon-ditioned
Up to 5	0.05	0.06	0.15	0.02
5–9	0.075	0.10	0.18	0.04
9–12	0.09	0.15	0.23	0.06
12–17	0.12	0.18	0.30	0.08
17–20	0.14	0.20	*	0.10
over 20	0.20	*	*	0.12

* Recondition before quoting.

Unscheduled labor costs

From Form 18-3—total preventive maintenance contract bid (labor) $____

Labor cost factor selected from table below ×____

=$____

Projected annual cost of labor, both scheduled and unscheduled $____

Labor Cost Factors

Age of System, yr	Prior Maintenance: Good	Poor	None	If First Recon-ditioned
Up to 5	1.2	1.3	1.7	1.15
5–9	1.3	1.45	1.75	1.25
9–12	1.4	1.6	1.8	1.3
12–17	1.5	1.75	1.9	1.35
17–20	1.6	1.9	*	1.4
Over 20	1.7	*	*	1.5

* Recondition before quoting.

Total general maintenance contract bid $____

Component Replacement Costs

The first thing to recognize is that you don't need to have exact pricing, as you do for a new construction estimate. You have a little more latitude here in arriving at a total for the cost of control components. This is not to say that you don't need to have a good knowledge of your costs for devices. However, in pricing a general maintenance contract proposal, the idea is to arrive at a total cost that is close enough to the actual value so that the price you develop for anticipated annual replacement is reasonably correct. Remember, you have no way of knowing *for sure* what the cost of replacement equipment for the year will be, for you cannot know in advance exactly what will have to be replaced. However, you can estimate accurately, using methods that are designed to guide you toward as precise an approximation as can be obtained. Thus, rather than develop validated valve sizing (valves already exist) and determine costs of dampers (they are already in place), you will instead consider the potentially replaceable items and eliminate from consideration fixed equipment that does not really need to be replaced, such as thermostat rough-in fittings in the wall, air tanks for compressors, dampers, panel enclosures, and tubing and wire.

In the cost of controls section of Form 18-4, thermostats are the first items listed. Count wall-mounted and unit-mounted thermostats, both electric and pneumatic. When figuring your costs, use the average cost of various pneumatic and electric wall-mounted and unit-mounted thermostats. Don't include the price of rough-in equipment. By consolidating prices and averaging them, you can estimate the cost of system components more quickly. (Remember, you aren't going to have to go out and buy each one according to a predetermined usage and temperature range, as you would for a new job.) For the purposes of arriving at replacement costs, you can group terminal unit and reheat coil controls by unit and reheat coil, respectively, and enter the total amount. You can use the same approach for pneumatic and electronic temperature sensors, pneumatic and electronic receiver controllers, EPs, and PEs, pneumatic and electric switches, pneumatic and electric damper motors (actuators), and linkages. Your familiarity with control component pricing from bidding on new work and ordering parts will come into play, and you'll quickly find that many of your prices will come from memory.

Next, count the smaller control valves, those 2 in and under, and enter an average price for each; do the same for valves 2 1/2 to 4 in and for valves 4 in and over. Valves that are 2 in and under are usually threaded, and those that are 2 1/2 in and over are usually flanged. This difference allows you to categorize them and use unitized average pricing for each of the three different size ranges of valves. You might find that the average price of a control valve from 1/2 to 2 in is $215.00, but you might choose to bias this up or down, depending on whether there are greater numbers of smaller or larger valves in this particular grouping. For 2 1/2- to 4-in valves, you might find that the average price is $650.00, but here too you might want to increase or decrease your estimate, depending on whether there are greater quantities of smaller or larger valves. The cost of replacement packing kits is covered in the estimate for the valve, and therefore no entry is required for packing kits. One point to keep in mind is that since you are not actually planning to purchase each item on your estimate, the costs that you estimate can be average costs. Your goal is to arrive at an approximate total cost in as efficient a manner as possible.

"Gauges" is a category that can cover a lot of territory, but none of the gauges envisioned here are terribly expensive. The gauges here range from those 1 1/2-in-diameter, 0- to 30-lb, 1/8-in threaded, back-connected "cheapies" to 3 1/2-in-diameter panel-mounted receiver gauges calibrated in degrees Fahrenheit. An average price of $20.00 to $50.00 might be appropriate, depending on the mix.

Notice that only the pump and motor of the air compressor are considered here. After all, the tank should last indefinitely. The pump and motor are priced as an integral unit, as are the air dryer, pressure-reducing valve, and filter set.

Pneumatic and electric remote-bulb thermostats have to be counted, and they can be grouped into one major category of pneumatic remote-bulb thermostats and one major category of electric remote-bulb thermostats. A remote-bulb thermostat does not sense temperature internally; instead, the sensing portion of the thermostat is connected to the actuating portion through a capillary anywhere from 5 to 50 ft in length.

Because they are expensive, enthalpy controllers and morning start-up optimizers are grouped together, and one average price is entered for them both.

Static pressure devices are also considered as a group and include high-limit and low-limit transmitters and controllers, all of which can be assigned a common denominator in terms of their cost. Firestats and freezestats are grouped together because they are both in the safety circuit of the fan motor starter and their costs can be averaged. They are not big-ticket items, and if one or two are missed in the count of the many firestats involved in fans throughout the building, it isn't critical to the estimate. The duct smoke detector is also in the safety circuit but it is more expensive than the previous two items and so it is listed separately.

You don't want to overlook the air and water flow switches, both the positive-displacement type and the differential type. Count how many you see, and especially watch for ones in condenser and chilled water lines; these switches may not have been included in the original temperature control contract but may be your responsibility in the general maintenance contract.

The time clock category includes those 24-h and 7-day spring-reserve or standard motor-driven adjustable-dial timers that have been prevalent in temperature control systems for years and are still very much in use today. The category also includes manually set wall- or panel-mounted spring-driven 0- to 12-h switch-type timers for overriding the action of a central clock during off-time hours. Interval timers fall into this category too. The average price here could range from $15.00 to $85.00, and would be influenced by whichever type of timer you'd need more of.

The "Industrial controllers" entry allows you to take into account any industrial-type equipment that was used for control where accuracy and the need for consistent close repeatability just couldn't be met with commercial equipment. On the next two blank lines you can list items not suitable for inclusion in any of the other categories. Finally, you can multiply quantity by cost to arrive at the total current estimated value of the system's control components. Now you have a financial picture of how many current-day dollars are tied up in the control devices you intend to maintain for 1 year.

This total estimated value is carried over to the controls annual replacement cost section of Form 18-4, where it is multiplied by the components cost factor, selected from the components cost factors table. The components cost factor takes into consideration the age of the system and the amount of prior maintenance that it has had. Note that the last column in the table is headed "If First Reconditioned." This means that if you are first allowed to overhaul a neglected system and get it to the point that it can be considered a worthy candidate for a maintenance contract, you can use the factors in this last column, which are much lower and allow for your reconditioning efforts. Once you have figured the total components cost, add overhead and profit to arrive at the projected annual cost of replacing components. This is the first of two totals to be added to arrive at your proposed total general maintenance contract bid.

Labor Costs

Labor needed for work in addition to semiannual preventive maintenance measures is covered in the unscheduled labor cost section of Form 18-4. You must plan for the total labor costs to be expected, which include planned preventive maintenance labor costs and the additional labor costs associated with trouble calls from your customer relating to component failure or the need for supplementary adjustments and diagnostic work. To develop unscheduled labor costs take the total preventive maintenance contract bid from Form 18-3 and enter it here. Multiply this figure by the appropriate labor cost factor selected from the labor cost factors table, which, like the components cost factors table, also takes into consideration age and prior maintenance of the system. In both tables asterisks indicate when reconditioning should be a prerequisite to bidding on a general maintenance contract. In the labor cost factors table, the preventive figure is always adjusted upward, but the degree of upward adjustment is what makes the difference. Again, system reconditioning comes into play, as the factors assigned here are the lowest. Use the appropriate factor to arrive at the projected annual cost of labor, both scheduled and unscheduled. Add this total to your projected annual components cost of replacements, and you have your total general maintenance contract bid.

In the unscheduled labor cost section, overhead and profit are not considered, because they have already been figured into the preventive maintenance contract bid (which is multiplied by the labor cost factor), and they rise together as a function of the labor factor applied.

Estimate your maintenance work, plan your work, and then work your plan. There is no substitute for carefully carrying out the maintenance plan you develop for any given system. Disregard your plan, and you will sustain cost overruns in the maintenance of the system. *You must not neglect your preventive maintenance and general maintenance contract customers!* Fail to service them regularly and effectively, and you won't keep them long. It is far better to be thorough in your preventive maintenance efforts and perhaps exceed your planned costs than to neglect your customer just because you are about to exceed your planned costs. If you have been thorough, you can expect your customer to accept an increase in contract price at renewal time. Your conscientiousness is apparent, your honesty shows, and your customer has been pleased. Be thorough! Be fair! Be profitable!

19

Retrofitting Older Temperature Control Systems

PLANNING FOR RETROFIT DESIGN

Existing temperature control systems should be examined with an eye toward improving efficiency and cutting operating costs. While most older systems which have been properly maintained still continue to function and control temperature, they do not always do it as economically as a modern installation that incorporates energy management techniques. However, older control systems can often be retrofitted with newer devices designed to lower costs and improve performance.

No retrofit work should be undertaken until an engineering study has been conducted. Such a study consists of an evaluation of the sequence of control events that occur on a rise in temperature and on a fall in temperature, along with an on-site inspection of the controls to determine their present state of maintenance. Additionally, the adaptability of the air-conditioning system to energy management devices is examined. Then, after a thorough system appraisal, retrofit plans can be designed. Existing control equipment, insofar as possible, should be reused wherever practical to incorporate minimum component cost with maximum return for retrofit dollar expended. This practice, along with a few others, is discussed next in the guidelines relating to designing for retrofit.

Retrofit Design Guidelines

Following these few guidelines for using new controls in energy management retrofit work will ensure the building owner of maximum economy and help you to win the job.

1. Reuse as many of the existing controls as possible. Base your bid on the controls to be reused, with a proviso that unusable controls will be replaced

at the owner's option at additional cost. Specify the terms for the cost-plus procedure to be followed when you encounter controls that have deteriorated to the point where they can no longer be used. This allows you to develop a quote for applying retrofit technology only, and the owner can more properly evaluate your submittal in light of what really needs to be corrected. Make sure the owner realizes that you cannot always tell which controls have to be replaced until work is under way, since it is frequently impossible to evaluate the condition of existing systems properly until the devices have been thoroughly checked out. In this way you can protect yourself when working to upgrade the controls for maximum performance and energy savings. Many systems suffer from improper maintenance even though they have been under maintenance contracts, often because the mechanic who performed the work for the service contractor lacked the incentive to do a thorough job. In addition, service contracts that are underestimated don't allow for mechanics to spend sufficient time to do a thorough job. "Putting out fires" may have been the tack taken under a tightly bid service contract to try to preserve contracts costs within reasonable bounds. You can see, then, the need for retrofitting contractors to protect themselves along with considering the performance of the system.

2. Purchase compatible devices. Oftentimes, manufacturers design different switchover points into such devices as pneumatic relays and changeover thermostats to limit interfacing with competitors' products, thereby hoping to preserve renewal and extension work for themselves. Be sure you know the operational characteristics of the existing components. Probably, the most feasible approach to successfully modify what is already installed is to select equipment offered by the manufacturer of the original controls. This should not be difficult since practically all manufacturer's controls are available over the counter now. Where simple systems make use of a single common main air supply to all devices, perhaps no compatibility problem will exist, but take the time to make sure. Since the circuitry for electric proportioning controls can vary from manufacturer to manufacturer, it may be necessary to select similar items from the original manufacturer in order to successfully modernize this part of the system. Manufacturers' catalogs and owners' maintenance and repair booklets can be extremely helpful. Such maintenance and repair booklets should have been delivered to the building owner at the completion of the original installation. When you are estimating the upgrading of a system, ask to see all the information available.

3. Determine the quality of the compressed air being fed to the pneumatic controls. Newer devices feature smaller ports and passageways, and clean, dry air is essential to their proper operation. Carefully check the air compressor to see how much oil carryover is making its way into the air lines where older oil-lubricated compressors are at work, how effectively the air dryer is removing moisture from the compressed air being fed to the

controls, and if oil or water is present in the sample devices removed for inspection. Problems such as these might not interfere with the operation of older controls because of their size and larger passageways, but in all likelihood, these problems will not be tolerated by newer pieces, which demand compressed air of high quality for good performance. Consider replacing the compressor pump and allow for this in your estimate—this could well prove to be an asset for both you and the owner. Also, allow for replacing filter elements and moisture traps if you have any doubt about their ability to do their job. Making such modest investments when the retrofit work is done adds immensely to the prospects for success in improving system performance.

4. Let the owner handle damper and valve body replacement. When new dampers and control valves are necessary, suggest that the owner hire others to perform the sheet metal work associated with damper replacement and the piping work associated with valve body replacement. Perhaps the owner has sufficient in-house talent to do this work. If not, your help in selecting tradeworkers to do the work will be appreciated. At least break out this work when your own mechanics are expertly qualified to handle it. This provides your owner-customer with options. If you prefer a purely controls contract, such options encourage your customer to permit you to do control work only, letting others come in to do the craft work that will support your activities. For years controls manufacturers have excluded damper and valve mounting work from their estimate, and quite naturally so, as their bids for the most part were to mechanical contractors, who would do it anyway. Leaving yourself an opening for avoiding this work can be of benefit to you and also to the owner, particularly if you are a controls specialist, because the heating and air-conditioning contractors might be reluctant to give away work which their laborers handle routinely on a job. But even so, a breakout for this work helps to establish that you recognize the costs pertinent to just the controls work.

5. Point out the advantages of polyethylene tubing, and base your bid on its use. Even when systems have been installed with copper lines, use of today's instrument-grade polyethylene tubing is far more practical and economical than replacing copper with copper. Preservation of the all-copper job is unnecessary, and the polyethylene tubing can be installed quite neatly, so as to reflect good workmanship. For runs to space sensors or thermostats that must be added, explain the accepted practice today of installing exposed single lines above suspended ceilings using proper supports. For fan rooms, explain that single-sheathed or multiple-sheathed polyethylene lines run to true horizontals and verticals can be made quite pleasing to the eye and will serve just as adequately as copper tubing where danger from high temperatures or mechanical damage is not a problem. Then the owner will see that your use of polyethylene simply represents a modern approach and is not an attempt to install piping of lower quality.

Using polyethylene helps hold down costs for the owner and for you, and you are more likely to produce an estimate that the owner won't feel is excessive. The best possible job for the lowest possible cost becomes the common objective of both you and the owner.

6. Reuse existing tubing when possible. Insignificant as it may seem, the reuse of tubing in renovation work can mean real savings for you in labor costs. Visit the job to determine how tubing is presently installed. Such an approach can reveal cost savings which can be brought into the estimate to produce a competitive edge at bid time. Once you have decided to reuse the tubing, you must make sure that the tubing will not be removed during demolition. This calls for careful coordination between you and the demolition contractor. Those above-ceiling lines which are to be preserved can be identified with suitable tagging or marking so that demolition personnel are alerted to keep "hands off." This time spent in coordinating and marking is necessary to protect your estimate, which takes into account the existing tubing. Once identified, each existing run should be clearly shown on your tubing layout for the temperature control workers.

   Any effective means of identifying the tubes you want to keep can be used. One method is to use the same markings that the general contractor uses; this way the demolition people will readily recognize that the items so marked are to remain undisturbed. Marking the tubes with red paint is another good way to indicate the tubes that need to be left alone. In addition, to protect these tubes, it is also important for your fitter to begin work before the demolition work starts. You may want your fitter to cut lines at strategic locations and roll back loose ends to clear areas where you know that dismantling activities are likely to interfere with the pneumatic lines already in place. Furthermore, the fact that the fitter is on the job early strengthens your request to leave certain tubes alone. This way, others will see your salvage efforts early in the game and know that you will be on the job to check on the control work. This increases the chances that the tubing you plan to use will remain intact.

When estimating retrofit work, it is essential to make adequate redesign plans, and the points just discussed should be taken into consideration each time you survey an older job for retrofit. Practical approaches that include retaining existing devices help keep costs down and are of great interest to the building owner. In the following pages, some typical retrofit and modernization applications are examined. In each case, the retrofit is based on (1) the desire to reduce energy costs and increase system efficiencies and/or (2) the need to upgrade areas to higher levels of comfort control as a result of changes in occupant activity.

Systems that are candidates for retrofit are wasteful of energy, not necessarily because they were poorly designed, but rather because at the time they were designed, there was little emphasis on energy efficiency. The design embodied in these older systems is usually adequate but not oriented toward conservation of energy costs. Sometimes, changing just one ingredient in the control scheme can enhance operational efficiencies significantly, particularly when a scheme is repeated many times throughout a building. Such is the case with two-temperature dual-set-point free-energy-band thermostats, which are ideally suited to dual-coil fan coil units and VAV boxes with reheat coils. In the past, the heating and cooling functions of these terminal units were separated by no more than a 1 or 2°F change at the wall thermostat. By applying today's energy-saving pneumatic thermostats that feature separate branch lines, separation of heating and cooling functions can be adjusted over a range of up to 20°F. Sometimes, occupants may have difficulty in accepting more widely fluctuating temperatures, especially if the new temperatures are introduced suddenly and the differential is set wide to begin with. It is frequently advisable to use dead-front nonadjustable (except by a special key) energy-saving thermostats and gradually, consistently, but slowly, widen the differential.

CONVERTING TO PNEUMATIC NO-ENERGY-BAND THERMOSTATS

For two-valve fan coil units or valve and damper motor VAV boxes, an additional pneumatic branch line generally needs to be installed from each thermostat to the terminal unit being controlled. A two-tube pneumatic cable was usually employed in the original installation; where this is so, the cable can often be used as the pulling lead to pull in three-tube cable, and then the old two-tube cable can be abandoned. Thermostat fittings which facilitate wall mounting will undoubtedly need to be replaced also.

Figure 19-1 depicts a VAV box with a hot water reheat coil. Figure 19-1*a* shows the piping of the original thermostat, and Fig. 19-1*b* shows the piping of the new one. Split-spring ranges between valve V-1 and damper motor M-1 exist, because in the initial installation the common thermostat branch line signal went to both of them, and the thermostat could first position the valve for coil bypass on a rise in temperature before opening the VAV box primary air damper for greater primary air for cooling. In doing the retrofit, only the thermostat was changed and an additional line added. The heating valve is controlled now (Fig. 19-1*b*) by the 65°F heating set point of the heating portion of the new dual-temperature thermostat, above which temperature (adjustable) no reheat occurs. The primary air damper is controlled now by the 78°F cooling set point of the cooling portion of the new thermostat, below which no cooling occurs. Identical spring ranges could be used in a new application, of course, but the existing split ranges do not cause any noticeable widening of the predetermined energy-free deadband. Room temperature may float between 65 and 78°F without energy being expended for heating or cooling. When this type of thermostat is broadly applied within a building, primary air requirements are lower and less hot water is needed for reheat.

Base your estimate for this segment of work on the typical number of thermostats involved. Determine the cost of the thermostats, as well as the

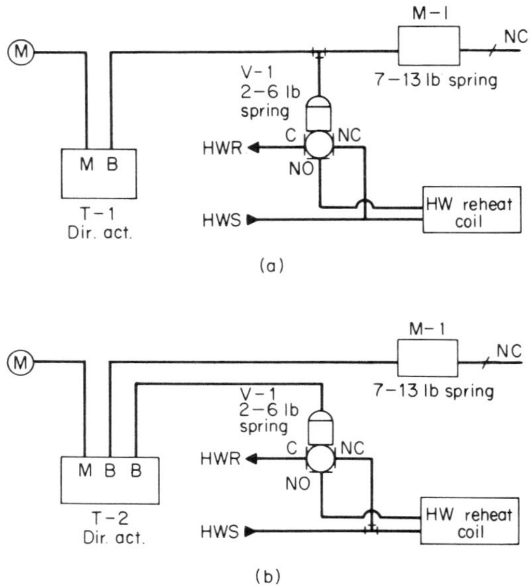

Figure 19-1 Retrofitting control of VAV box with reheat. (*a*) Original control; (*b*) energy-saving thermostat.

mounting hardware and tubing costs for each, and make an average allowance of tubing for each terminal unit if all the runs can be averaged and are a similar length. Allow about 2 labor-hours for each thermostat.

Caution! Make sure that the air compressor can handle the additional compressed air requirements of the new thermostats. Dual-set-point energy-saving models may use more air than conventional models, and if many are involved, you'll need to note the run time of the compressor and calculate the effect of increased air consumption. The thermostat manufacturer's data sheets will list the device's air consumption, and the compressor run time can be checked by timing the on and off cycles to make sure that the additional air consumption will not result in excessive running of the compressor. You can determine the compressor pump output capacity by noting the compressor's model number and calling the local representative for the manufacturer.

Occasionally, constant-volume boxes with reheat coils are retrofitted using kits that introduce VAV capabilities into the old box. When this is the case, automatic actuators, usually pneumatic, are added to the box by using an air valve extension kit, which allows a volume regulator type of air valve to be incorporated into the older unit. You may or may not have to mount the actuator, depending on whether the manufacturer factory-mounts the actuator before the retrofit kits are shipped to the mechanical contractors. If you are going to do the mounting, use the labor-hours for a damper motor given in

Table 3-1, which lists the equipment room labor-hours for control devices; the time allowed for 30 ft of tubing provides for the extra branch line that would be required if the retrofit work included installation of a three-pipe deadband thermostat.

Caution! Often the work is in existing finished areas, and this means that the ceiling materials may be in place. Be sure to allow time for removal and replacement of the ceiling tiles. If the ceiling is splined, it is likely that just certain locations in the ceiling will permit access to the boxes. The same is probably true for a hard-cast ceiling, where panels built into the ceiling provide the only access. You must be sure that you can do your work through the openings that are available. If not, you'll have to make special arrangements with the owner for providing access. Carefully check the situation and satisfy yourself that you can reach all the areas you need to get to. When it is necessary to remove strategic portions of a hard-cast ceiling so that you can get to the work areas, the building manager might elect to use in-house maintenance people to provide the openings for you. The willingness of the building owners in going this far will depend on how much they wish to cut operational costs, which probably will depend on just how expensive the building has been to operate from a heating and air-conditioning standpoint. Discuss all the facets of the work involved, and help the owners to arrive at a satisfactory decision.

VENTILATION LOCKOUT WITH ENTHALPY CONTROLLERS

Energy management devices can be applied in existing older systems in several areas, depending on the equipment utilized in the building under consideration. Space temperature controls, fan systems, and heating-cooling power generation equipment all are candidates for modernization. Let us look now at the air-handling unit and see what might be accomplished to improve efficiency.

"Enthalpy" means the total heat content. In determining the enthalpy of air, you must consider both the temperature and the relative humidity of the air. Once you know these measurements, you can refer to a psychrometric chart to find the enthalpy. To reduce the cost of mechanical cooling, you want to select the air for use in the air handler that has the lowest enthalpy when the chilled water or direct-expansion cooling coil is in operation. Consider a spring or fall day when a conventional changeover thermostat sensing outside air temperature has given the signal to close off the fresh air and start the mechanical cooling equipment. Ordinarily, when the cooling coil begins to operate, the fresh air dampers of the air handlers return to their minimum open position to reduce the load that would otherwise be created by warm, humid air from the outside passing through the unit. However, the outside air enthalpy might be lower than the enthalpy of the return air coming to the unit from the building. This can well occur if, as is often the case when people and activities are involved, the relative humidity of the return air taken together with its temperature produces a higher enthalpy than exists in the fresh air being drawn in from the outside. In such a case, it is more economical to cool down the outside air than it is to cool down the return air. An enthalpy control employed at the air handler to govern the introduction of fresh air and return air into the unit could mean energy savings for the building management.

In the past, decisions about using outside air for cooling were made purely on the basis of temperature; they did not take into account the adverse effects that high relative humidity can have on the efficiency of a unit in the cooling mode. Previous changeover applications often involved an outside air thermostat set at 60°F. Above this set point, dampers returned to the minimum fresh air position, and mechanical cooling was started. Enthalpy control continuously monitors the temperature and relative humidity of the fresh air and the return air. The enthalpy, or total heat, of the air is a better measurement of the air's actual heat content, the total heat that a cooling coil must deal with. In actuality, even though the temperature of the outside air might be higher than 60°F, the outside air, in comparison with the return air coming to the unit from the spaces served, still represents the lesser of the two heat loads to be handled by the air handler. In a typical retrofit application, the outside air thermostat might be replaced by an enthalpy controller with temperature and relative humidity sensors in the outside and return air. Figure 19-2 shows this kind of retrofit application. Note in Fig. 19-2*a* that the traditional remote-bulb thermostat, HL-1, is located to sense the outside air temperature and to lock

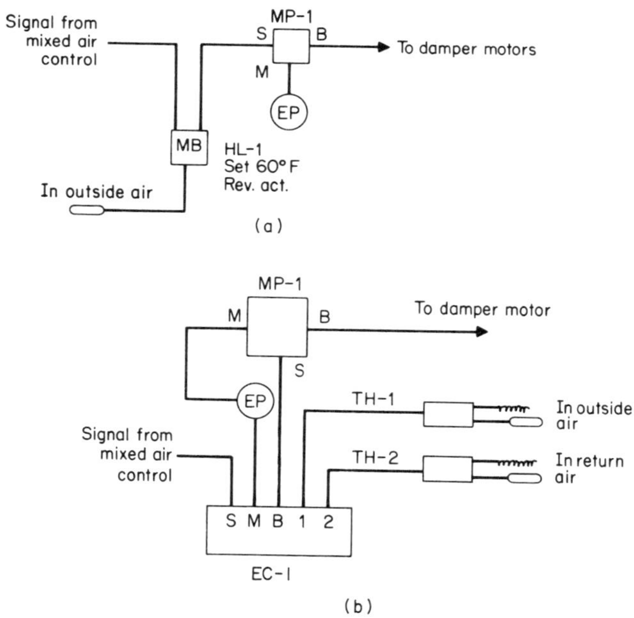

Figure 19-2 Enthalphy retrofit of outside air changover. (*a*) Original control; (*b*) retrofit control.

out the signal from the mixed air control above 60°F so that the ventilation cycle is inhibited. HL-1 removes signal air from minimum position switch MP-1 so that the dampers will assume their preset minimum position. In Fig. 19-2*b*, enthalpy controller EC-1 has been installed in place of HL-1, and combination temperature and humidity sensors TH-1 and TH-2 transmit conditions of the outside air and return air to EC-1. (Some manufacturers use separate temperature and humidity sensors.) Now the ventilation signal from the mixed air controller will be inhibited only if the enthalpy, or total heat, of the outside air is greater than the enthalpy of the return air. It is assumed in this illustration that start-up of the mechanical cooling equipment is handled separately. EP main is removed during unoccupied periods for positive closing of the fresh air dampers.

The work associated with this retrofit consists of removing HL-1, installing EC-1 with its associated transmitters TH-1 and TH-2, and doing the necessary piping, all of which will most likely be done within the mechanical equipment room. In estimating this work, price the controls, measure the tubing required, and allow sufficient labor-hours to make the change, including calibration and setup. For a retrofit like this, you could also estimate the devices as new work within the equipment room and use normal fan room tubing allowances.

MORNING START-UP OPTIMIZATION

The operation of an air-handling unit can also be greatly improved through modifications relating to morning start-up and the lockout of outside air during the warm-up period. Morning start-up optimizers, now offered by virtually all commercial controls manufacturers, shift the restart time of the air handler during the heating season so that as the outside temperature rises in the early spring, the start-up time is delayed, on the theory that milder outside temperatures make it easier for the unit to reheat the building to daytime temperatures before the occupants arrive. Additionally, the introduction of fresh air is delayed until after warm-up is completed, and so the recovery time is minimized. In the past, a time clock was employed for starting and stopping the unit at predetermined times. Once set, these times governed the unit's on and off operation and could be changed only by intentional manual intervention. With the introduction of the morning start-up optimizer, varying lengths of warm-up time can be scheduled to correspond automatically with the building's need for warm-up, as determined by different sequencing cams which are biased a proportionate amount as the outside air temperature changes. By matching the building coefficient of heat loss with the closest program cam, a curve can be established which delays the on time in response to warmer outside air temperatures. Although now regularly used in new installations, the morning start-up optimizer is rarely seen in older work, and therefore many older systems are candidates for this particular retrofit.

Traditionally, shutdown of the air handler during unoccupied periods is coupled with lockout of outside air, and the ventilation cycle is reinstituted upon daytime indexing of the equipment. Figure 19-3 shows a typical time clock and EP night ventilation lockout control scheme before retrofit. Contacts A and B of the clock activate simultaneously and are closed during the occupied period. Such action energizes EP-1 and also fan motor starter FMS-1. This starts the fan of the air handler and allows the signal from the ventilation controller to pass to the damper motors. In this scheme, start-up occurs at the same time every day, and fresh air is introduced at once.

Restarting the air handler in the morning at the same time that the discharge thermostat resumes control of the dampers has caused more than one air-handling unit to lock out from freezestat actuation in cold weather. Because of shutdown, the discharge controller may be warmer than it would be during daytime operation, when it can control the temperature of the discharge air coming from the air handler and maintain its set point. As a result, when the air-handling unit starts up and the discharge controller is placed in command, the controller's first act is to push open the fresh air damper. Then the freezestat, which is often in the mixed air, gets cold and reacts by opening the electric circuit to the fan motor starter, shutting down the fan before the discharge controller is allowed to cool down and take control of the dampers. One of the advantages of the morning start-up optimizer is that it protects against sudden opening of the fresh air damper by using a two-stage process to restore normal daytime temperatures gradually.

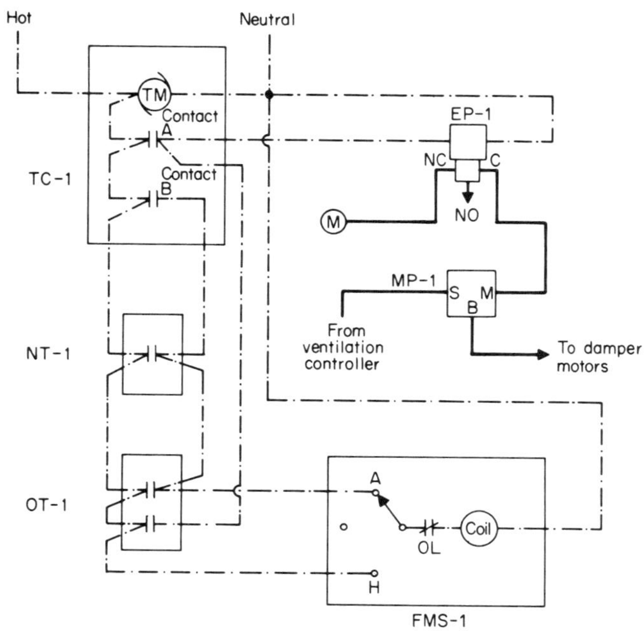

Figure 19-3 Conventional occupied-unoccupied controls.

Figure 19-4 shows the lockout control scheme from Fig. 19-3 after the retrofit. Note that morning start-up optimizer MSO-1 takes the place of the conventional time clock. Also note that pneumatic temperature transmitter TT-1 has been added, which signals the pneumatic biasing module within MSO-1 to indicate how cold it is outside and how early, therefore, start-up must be accomplished to restore daytime temperatures. Sequentially, MSO-1 is indexed by the outside air temperature whenever the air is colder outside to start the HVAC equipment earlier. First contact A closes, in order to start and run the fan of the air handler until normal occupancy time. Then contact B closes, allowing the introduction of the minimum amount of fresh air required by codes for occupied buildings. At this time, the unit is out of the warm-up mode and into the normal daytime mode of continuous fan operation, with at least the minimum amount of fresh air being drawn into the building. Later on in the evening, as determined by building management, both the contacts open, the fan operation ceases, and the dampers close off to outside air. Two-position hand-off-auto (HOA) selector switches S-1 and S-2, which are built into MSO-1 and are accessible to authorized personnel only, allow the overriding of contacts A and B as may be necessary for testing or for

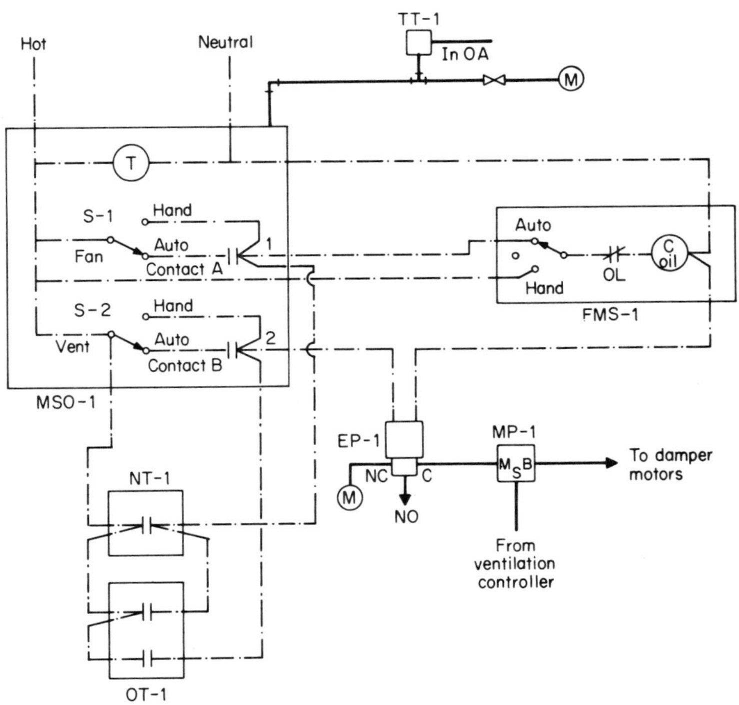

Figure 19-4 Retrofitted occupied-unoccupied controls with morning start-up optimizer.

assuming manual control. Night thermostat NT-1 brings the unit back on at night if interior temperatures fall to its low limit. And double-pole override timer OT-1, which is accessible to occupants, allows the occupied mode of control to continue beyond normal shutdown time at night as may be necessary for after-hours activities.

This retrofit can be estimated on the basis of fan room work if the original time clock is already in the equipment room. TT-1 can most likely be mounted on the outside of the fresh air damper, which also makes it suitable to be estimated as fan room work. A new pneumatic line must be run to TT-1, but no new wiring conduits should be necessary. To estimate this job, enter MSO-1 and TT-1 on the equipment room labor and materials estimate (Form 3-1). Select the appropriate labor-hours from Table 3-1, which lists the equipment room labor-hours. TT-1 qualifies as a rigid-stem sensor, and so when you locate this row in the table, you can select the correct hours for the tubing you will use. MSO-1 qualifies as a receiver controller because it uses the linear pneumatic input from TT-1 to determine the length of the warm-up cycle. It will also be necessary to remove the old time clock and reconnect its wires. To allow for this, enter the time clock on Form 3-1 without the cost. Instead, merely enter the electrical labor-hours from Table 3-1 to cover disconnection, dismounting, and removal, and such modifications to the conduit system at the time clock location as will be necessary, plus reconnection of the wires. This "labor-only" entry ensures that time will be allowed for performing the electrical work incidental to this retrofit. No entry for wiring materials should be necessary. Price the retrofit accordingly and then fill out the bid summary sheet (Form 14-1).

The standardized estimate forms are well suited for figuring such work as this. Use them in conjunction with jobsite visits that serve to familiarize you with the actual job conditions. You can adjust your estimate to allow for the difficulties you have viewed at the jobsite and reevaluate the hours you have assigned for accomplishing the work.

Labor has often been viewed as the biggest variable on a construction project. Keeping the cost of labor within the bounds established by the estimator usually means that a good job is in the making. There is no substitute for your personal experience in appraising a job and the variables with which your installers will have to deal. It is a good idea to prepare retrofit diagrams at the time you visit the work area. These rough sketches will show the new devices that will be required and the existing ones that will be reused.

REZONING A SINGLE-ZONE AIR HANDLER

During a retrofit in which pneumatic controls are installed, a packaged air-conditioning unit which formerly provided heating and cooling in response to a single thermostat might well be modified by spec so that the reheat coils under the control of separate thermostats can temper air as required in individual rooms. In the case of a direct-expansion cooling coil and electric two-stage strip heat, a two-step heating, two-step cooling low-voltage thermostat could well have been the single control which indexed the unit for either heating or cooling and which cycled the fan as required to coincide with the demand for space temperature changes. When units of significant size serving an open area were involved, such a method was probably okay. Assume, however, that partitions will be erected in the formerly open area and that the individual offices will be heated and cooled independently of each other. Additional ductwork will be added to subdivide the air delivery off the air-conditioning unit, and hot water reheat coils will be inserted in branches to the individual offices. Each thermostat will modulate hot water control valves at the reheat coils for tempering air in the duct to its area.

Preferably, the refrigeration compressor will provide cooling only if an area requires it. The approach here will be to delete the existing heating-cooling electric thermostat and instead to allow the refrigeration compressor to run only if one of the individual thermostats calls for cooling. In looking at the existing unit, you note the following requirements which you must accommodate by careful application of controls:

1. The compressor must not be allowed to run unless the evaporator fan is in operation.
2. The electric heating coil in the unit also must not be allowed to run unless the evaporator fan is in operation.
3. Discharge temperature control of the direct expansion cooling coil should be avoided because of a tendency of this mode of control to create short cycling of the compressor.
4. Also, provisions must be made for fan control, because fan cycling occurred previously as a function of a need for heating or cooling.

Figure 19-5 shows pneumatic controls that have been adapted to a self-contained air-conditioning unit with a fixed, minimum amount of outside air. A direct-expansion refrigerant coil is used for mechanical cooling, and an electric strip heating coil is used for mechanical heating. In the unit as it was originally installed, the heating coil was in the reheat position, but for retrofit of the unit's controls, the electric heating coil is moved to the preheat position, and for good reason. Previously all space heating was accomplished by the electric heating coil, and its position didn't matter. Now the only purpose of the electric heating coil is to temper incoming air in extremely cold weather, as reheating will be handled by the hot water reheat coils on a room-by-room basis. Since temperature controller TC-1 will be set to maintain 50°F, a position after the cooling coil in the classic reheat position will cause the heating coil to operate when the unit is activating the refrigerant compressor for cooling. EP-1 will be wired through an auxiliary contact of the fan motor starter to pass main air to TC-1 and to selector relay SR-1. This ensures removal of pneumatic signal from these two devices when the fan is off so that the normally open PEs

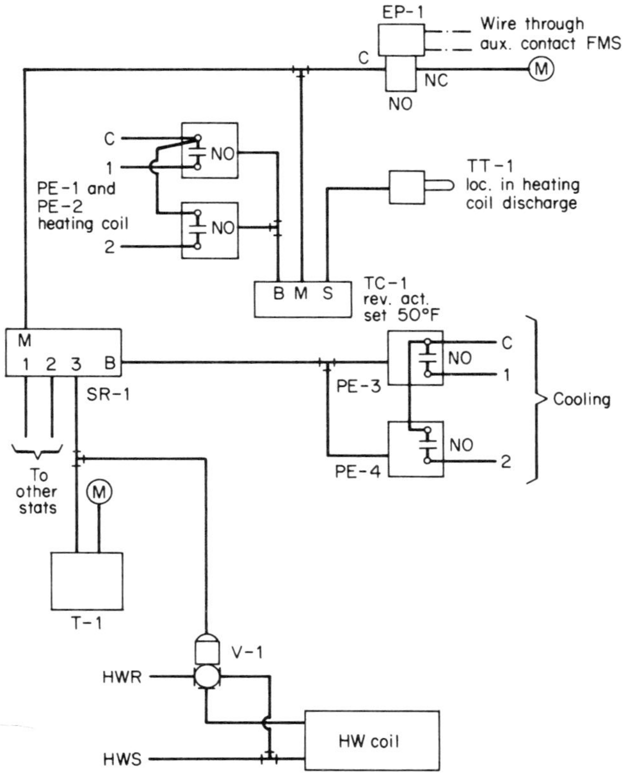

Figure 19-5 Pneumatic controls applied to single-zone packaged unit.

can open their contacts to prevent either heating coil or refrigerant compressor operation. Space thermostat T-1 (two others will be used but aren't shown in Fig. 19-5) will modulate its hot water coil reheating control valve to maintain space temperatures in the office. If a space thermostat senses space temperatures above its set point, it will position its control valve V-1 for hot water coil bypass and will signal SR-1 to pass its pneumatic branch line pressure onto PE-3 and PE-4 for operation of the compressor. The compressor will continue to operate, under its built-in coil capacity regulators, as long as a need for cooling exists.

USING EXISTING CONDUITS

Typically, there is much that can be reused in retrofit installations. One of the most obvious items that can be reused is conduit. Because of improvements in data retrieval and data gathering techniques, those systems with HVAC automation conduit frequently lend themselves nicely to the installation of new wiring to new sensors.

Consider a building constructed in the late sixties. Its automation system may have contained thermocouple temperature-sensing devices for readout on a central meter at the console of the automation system, which was then considered quite modern. Perhaps through matrix selection, the points within the system were indexed individually over the same groups of wires for time-shared used of the central trunk cable. Although these matrix-type, time-shared systems were an improvement over the older, hard-wired systems, where every point was individually wired back to a central display point, they still required designated wiring to each device for retrieval of information and issuance of commands, even though the wiring was connected selectively through point selection relays to the trunk cable going back to the central console. Selective coding, which permits a pair of wires or a coaxial cable to carry the signals from field data retrieval panels to the central console, wasn't introduced until years later. Each data retrieval panel would respond only to specific electronic queuing to which it alone was attuned. Other data retrieval panels had different electronic addresses and would not "answer" when signals other than theirs were on the common "bus." The next improvement was the development of field-located, computer-based data acquisition panels, which were capable of making intelligent decisions related to the HVAC process and contained preprogrammed information related to alarm limits, load shedding, and duty cycling. These lower-hierarchy computers could communicate with one another and with the central master processor at the command center. But no matter how simple or technologically advanced the devices were, they all required one common ingredient—conduit. The conduit system is an ingredient which can be salvaged from these older systems and reused. This is fortunate since the conduit system is one of the most expensive ingredients in an automation system because of the labor and expense involved in getting it installed. Having the system already in place creates savings and necessitates only the pulling in of new wiring that may be required by the system being installed. As new sensing devices are installed which make use of newer temperature-sensing and transmission technology, the old wire can be used as a pull wire—to pull in the new wire needed for the upgrade. When you estimate a retrofit installation involving an energy monitoring and management system, you need to look closely at the components of the old system that can be reused in the new one.

You want to be careful not to oversimplify matters just because you discover as you do your survey that the conduits you need are already in place. Consider their location and accessibility. Productivity drops in hot, noisy boiler rooms, and invariably conduits are run upward from pipeline devices to the poured slab or bar joist above, across the ceiling of the boiler room, through junction boxes, and on to the panel located in the boiler room for signal gathering, acquisition, simulation, and transmission. When pulling out the old wiring, plan on having to get up into the high portions of the boiler room where the junction boxes are. You will discover that these areas are hot and dirty, and sometimes you will find that the jackets of old transmission wiring have softened over the years in the hot boiler room and have adhered to the interior of the conduits, making pulling out the old wiring more of a challenge than you might have first envisioned. If you have already done this work yourself, you should have a feel for working under these difficult conditions. However, if you have not, you need to remember that even new work in an area such as a hot, dirty boiler room is not enjoyable work—certainly it is not preferred work from the viewpoint of the installer. You must make proper allowances for tackling this type of work.

When planning on reusing existing conduits, keep the following questions in mind:

1. What is the state of the wiring within the conduit which you will have to remove?
2. How accessible are the junction and pull boxes? Can they be reached with stepladders, or will special scaffolding be required to span the equipment in place? Over the years, changes can occur in the allotment of mechanical equipment room floor space as building maintenance and renovation occur.
3. Can the work be more economically and safely performed when the heating equipment is down, such as in the summer? How urgent is the need for immediate retrofitting?
4. Is the boiler room ever really cool? Do the boilers operate year-round because of absorption chillers and/or space reheat coils?
5. Can normal straight time be figured? Must any of the work be done after hours in order not to interfere with the existing automation system, which must remain up? It's rare that you will be required to make a change in an automation system that is running, but if control and alarm points are activated by out-of-range temperature conditions that may be simulated because of your disconnecting the equipment, then you must plan for daily interfacing of the new components for maximum uptime for the system.
6. How accessible are the areas in which you will be working? How much downtime will you incur because you have to go for keys to enter locked areas? How essential is security to each of the equipment areas in which your work will take place?
7. What is the time frame for normal daily activity in the building? Can you start at your normal time, or will the job necessitate different start and stop times?
8. Will any 120-V wiring in the existing conduits have to be rerouted to keep it from causing interference in the new electronic wiring?
9. Will all the wiring in the conduits be removed? If not, is there room to pull the new wiring past the wiring you must leave in place?
10. How mechanically hazardous is it beneath the area in which you will be removing old wiring? Pulling in new wiring invariably means pulling from

box to box or from box to fitting, and this results in long loops of wiring hanging from the ceiling. If hazardous motion machinery is in operation below the area in which you will be working, you will have to exercise extreme caution in performing your work, and this will affect your labor costs.

11. Is the retrofit work clearly defined? Your purchase order should identify the complete scope of your work and exclude areas for which you are not responsible.
12. Is the system in good working order? Before you begin work, you should come to some agreement with the owner about the condition of the section(s) of the old system that will remain untouched—for example, a fire alarm section that is part of the central installation, or an intercom module that is not going to be taken out, or perhaps a security system that is interfaced with the existing system and that is also to be interfaced with the new, retrofitted system. Unless you examine the portions that are to remain and point out the problems ahead of time, you might be blamed for failures that existed before you came on the job. You can see the need for careful examination of any system about to undergo retrofit, because these and other related situations can create undesirable misunderstandings between you and your customer unless they are controlled at the outset.

A system layout and riser diagram are needed for retrofit work, just as they are for new work. The system to be installed should be clearly defined in your own mind, and diagrams help best to accomplish this. Then, you can view the drawings against the existing raceways to determine how much of the raceway you can use in the new work.

Labor requirements for retrofit work may be more flexible in terms of timing, but probably no fewer than two workers should be assigned the task of pulling new wire. As with new work, it takes at least two workers to handle the wire, one pulling the wire and the other guiding it into the raceway without knotting and snagging.

OCCUPANCY INDEXING RETROFIT OF A ROOM UNIT VENTILATOR

In many heating-cooling unit ventilator installations, setback during the winter is accomplished by a central time clock with central or zone night thermostat low-limit protection. Once the unit ventilators are shut down by the time clock, zone thermostats set for reduced night temperatures send a pneumatic signal to PEs within the unit ventilators to restart them for cycling as unit heaters to maintain reduced temperatures during unoccupied periods. The room thermostat controlling the face and bypass dampers of the unit ventilator will, of course, call for heating, since it has a higher setting than the zone night thermostat. Thus, restarting the unit ventilator will cause it to deliver heat.

A further refinement to permit energy conservation is to integrate the operation of the room unit with the room light switch; when the lights are off, the room is assumed to be unoccupied, and when the lights are on, the room is considered to be occupied. Note in Fig. 19-6 that the light switch is used to activate EP-2 and, through air flow switch AFS-1, also to activate EP-1. The retrofit work will consist of installing two wires from the closest light or the light switch (if the electric neutral conductor is present in either the lighting outlet or switch outlet) to the unit ventilator. Most likely a clock circuit will already be in place from a central time clock to EP-2. In this retrofit, the clock circuit will be replaced by the extension of the light switch circuit into the unit.

When the light switch is indexed to turn on the room lights, EP-2 is energized to place main air on PE-1 for constant fan operation. When the fan is running, as sensed by AFS-1, EP-1 is energized, and the outside air and return air dampers are allowed to operate.

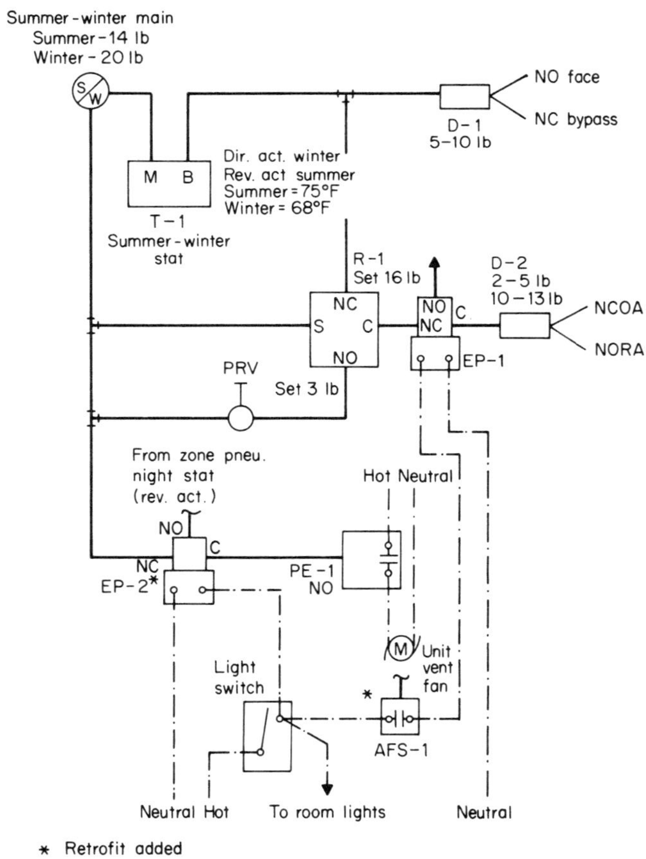

Figure 19-6 Heating-cooling unit ventilator occupancy retrofit.

During the heating season, 20-lb main air pressure indexes summer-winter room thermostat T-1 for direct action so that on a rise in room temperature, damper actuator D-2 gradually opens the outside air damper to its minimum position while positioning the return air damper closed a proportional amount. On a continued rise in room temperature, face and bypass damper actuator D-1 is modulated to decrease the amount of air passing over the heating coil while correspondingly increasing the amount allowed to bypass the heating coil. If the room temperature continues to rise, and after the face and bypass damper is in the full bypass position, the outside air damper is modulated up to 100 percent open while the return air damper is modulated 100 percent closed in order to provide for the introduction of outside air for cooling. On a fall in room temperature, the sequence is reversed.

If the room is not in use, the light switch should be turned off. When this happens, EP-1 and EP-2 are deenergized. EP-1 closes the outside air damper completely while opening the return air damper. EP-2 places PE-1 (which controls the fan) under the control of the zone night thermostat, but no signal comes from the night thermostat at this time because the building is warmer than its set point and its reverse action has its branch line at zero. Therefore, the unit ventilator fan stops. If the room temperature falls during the day while the room is in an unoccupied mode, the unit ventilator room thermostat opens the face and bypass damper to permit convection to take place, but the fan does not run. Since the unit ventilator acts as a convector during this period, unreasonably cool temperatures are prevented in the unoccupied room. As soon as the light switch is turned on again, the fan will start, and the room temperature will return to the set point of T-1.

During the cooling season, 14-lb main air pressure indexes T-1 for reverse action, so that on a rise in room temperature, D-1 is allowed to open for air flow across the coil, which now has chilled water flowing through it, for cooling. D-2 is permitted to open for a small amount of outside air only, because changeover relay R-1, indexed by the lower main air signal, switches to place pressure regulating valve PRV, set at 3 lb, in control of D-2, provided the room is occupied (the light switch is on) and the fan is in operation (AFS-1 is made). In the summer when the room is unoccupied (the light switch is off), the fan stops, and it will not restart at night because there is not a sufficient temperature fall in at the night thermostat. Thus, in the summer, the only time the unit runs is when the room lights are on.

In the original scheme, before the retrofit work was begun, an evaluation of the system would probably have revealed that the pneumatic signal from the night thermostat was piped directly to PE-1 and that another EP (not shown in Fig. 19-6), at the central time clock point, was used to bypass this night thermostat line during the day so that the unit ventilator fans would stay on. In the retrofit, however, EP-2 is added, and the night thermostat line is routed through the normally open port of EP-2, while local main air is introduced into the normally closed port. Back at the time clock, the piping is changed so that the EP at the central time clock point is no longer involved and only the night thermostat can now send a signal through this existing pneumatic line. The light switch is an existing item, but the best point has to be found to pick up the neutral along with the switched leg from the light switch and to extend two wires over into the unit ventilator. EP-1 is an existing item and originally was probably wired in parallel to the fan motor with one of its legs broken through another PE in the unit (not shown), which was probably piped back to a clock-controlled EP at the central or zone time clock point. It is rewired through AFS-1 to the room light switch. If the existing EP-1 is 120 V ac and if the light circuit is 277 V ac for a fluorescent light, the existing EP-1 must be changed to match. Therefore EP-1 could be a new retrofit item if its voltage is not suitable for connection into the light circuit. The air flow switch will probably have to be added.

Other schemes that can be retrofitted use a direct-acting night thermostat. In these kinds of schemes, at a predetermined time in the morning, main air is simply removed from the night thermostat, and a normally closed PE in the unit ventilator makes the fan run constantly during the day. Then, at a predetermined time in the evening, main air is placed back on the night thermostat, which is warmer than the day thermostats (owing to its set point). It immediately passes air and fills its branch line, opening all the normally closed PEs to which it is connected in the unit ventilators.

The retrofit in this case consists of repiping the night thermostat so that it has main air on it at all times. During the day, its signal is sent to the unit ventilators, and a normally closed PE-1 is used. More likely than not, the original PE is equipped with single-pole double-throw contacts, so that mere transfer of the conductors to different terminals can achieve the normally closed contact configuration needed with the direct-acting night thermostat. With this in mind, EP-2 is piped without the main air connection to its normally closed port, and the port, instead, is vented to the atmosphere. The night thermostat continues to be piped into the normally open port of EP-2. Now, with PE-1 normally closed, turning off the lights returns EP-2 to normal, and the signal from the night thermostat comes through to open the contacts of PE-1. In Fig. 19-6, an asterisk indicates the items added during the retrofit. Pneumatic work is confined to unit ventilators, but wiring must be run to the light circuit.

In doing an estimate for the retrofit just described, you would need to include two new controls per unit ventilator, and to allow the necessary time to accomplish the retrofit. In addition, you would need to study the best route for tying into the switched leg off the light switch. Then, whether the time clock and night thermostats are zoned or are in a central location, you would have to estimate time for performing the modifications required for your particular system. Compared with installing a central energy management system, this particular retrofit would probably cost a lot less money and yet result in significant savings and provide greater flexibility to the building operators.

Wherever terminal units are used, look at the scheme employed. When the installation has a few years on it, it's a safe bet that you can find ways to save the owners money on the operational costs. Consider above-ceiling equipment as well as freestanding units, and examine each with an eye toward improving economy of operation. Frequently, the improved efficiency requires very little sacrifice in terms of occupant comfort. With large numbers of units, savings multiply rapidly when cost-cutting measures are introduced. Once you exercise your expertise in this area, you will enjoy a reputation for being well versed not only in installing new control systems but also in improving older ones. Remember, study the systems carefully first, and plan for maximum benefit for minimum cost.

20

The Automatic Temperature Controls Proposal

It is good business to back up your telephone quote with a written confirmation, for what you say over the phone can be misconstrued. A written proposal will allow you to confirm your temperature control quotes, and when properly designed, it can also serve as a means for obtaining a contract for the work. Probably, you've seen this kind of proposal in some of the standard forms that contractors offer. A proposal form that I use and recommend is presented here as Form 20-1. It contains the essential ingredients of a contract. The proposal states that I will provide a system of automatic temperature control (offer) for a stated amount of dollars (consideration). The offer is to provide a service which does not violate laws (legal consideration), and the parties to it, both the offeror and the acceptor, must be of age and of sound mind to enter into the contract (competent parties). In signing the proposal, the customer indicates approval of my offer and authorizes me to proceed (acceptance). Additionally, the project name, the date of the specifications, the specific plans, and the sections of specifications covered are included. Any addenda that have been made are acknowledged too. This is an important point, for if you have missed seeing an addendum, your indication of the ones covered may provide the basis for your obtaining an extra, if you are entitled to one. It is important for you to engage professional counsel in developing your own proposal form.

It's a good idea to itemize inclusions and exclusions pertaining to those questionable areas that you uncover in the process of making your takeoff. Such items might be mounting and wiring boiler controls, wiring water treatment equipment, wiring filter advance mechanisms, mounting smoke detectors, furnishing access doors, and doing patching and painting in existing buildings; the list can be endless, depending on the job and the situations you envision. Preprinted terms of payment indicate that you will be issuing monthly invoices for work completed and that they are to be considered as net amounts. Another good idea is to type the proposal in triplicate. Leave one copy in the job file, and take the other two with you at the time you expect to close the deal.

Harry J. Edwards, Jr.
CONTROLS CONTRACTOR
1009 Baldwin Road, Richmond, Virginia 23229
(804) 288-0657

Automatic Temperature Controls
P R O P O S A L

To:

Date:
Quote Firm For:

Job Name:

To provide a system of automatic temperature control in accordance with those specific portions of specifications listed below, and in accordance with the terms and conditions given below, for the total net sum of ______________________________(dollars)

$__________

Project Specifications Dated:

Specific plans, and specification sections covered:

Addendums Acknowledged:

Included in This Proposal

Not Included in This Proposal

Terms of Payment: Monthly invoices net as job progresses

Accepted:____________________ Proposed by:

By:____________________

Title:____________________

Date:____________________

____
Harry J. Edwards, Jr.
State Reg. 15448

Celebrating 20 years of Temperature Controls Experience in 1984

Form 20-1 Automatic temperature control proposal.

Leave one copy with your customer and keep the signed copy as proof of your contract. The unsigned file copy can be used to prepare duplicate copies if they are needed.

Obviously, Form 20-1 is best suited for work in which plans and specifications are involved, and this includes both new construction and rehab work. An architect is almost always involved in the work being done today, and so you can count on there being plans and specs for most of the jobs out for bids. Therefore, Form 20-1 can serve quite well. Of course, modify the form as you see fit, or devise your own form.

SELLING

After you've worked hard to estimate the job and then you quote the job and are low bidder, you're entitled to the work. Yet, 90 percent of the work I've booked over the years I've still had to ask for, which means that you have to be able to sell your work. Even though you're a contractor, selling is a very important part of your business. You need to ask for the order. I've used Form 20-1 over the years as a tool to obtain my customers' commitments. Some of them prefer to issue their own purchase order subcontracts later, but most of the time they're willing to give me an okay on the proposal form so that I can proceed with my work—getting the job booked, setting up a job file, planning the work, and preparing to issue purchase orders for the control devices and installation materials.

Know your customer and know when to ask for the order. Of course, before making a commitment, your customer must receive some assurance from the general contractor. However, since your proposal is simpler and lower in dollar costs than the mechanical contract, the general contractor may need more time than you do to get things together. If you're in a hurry to book the job, you can qualify your proposal by adding at the bottom of it, "This acceptance is subject to award of the work by the general contractor to the acceptor." This qualification will permit your customer (the mechanical contractor) to give you the okay you seek.

SCOPING THE JOB

Your written proposal also serves as a prenotification of what you will and will not be including in your bid. This is called "scoping the job." You want to keep your price confidential until the proper time on bid day when you usually phone it in to the mechanical contractor. However, oftentimes, particularly on a large job, the mechanical contractor appreciates having your proposal in advance of the bid date in order to be sure that you've looked at the job sufficiently to know exactly what's required of you. By submitting a proposal without a price on it, you can protect your price until the last minute, and your customer will know what to expect and can decide if the bidding temperature control contractors are all looking at apples and apples. In this kind of advance proposal, the word "scope" is added to the proposal, to make it clear that it is a scope quote. The scope quote is handy, and it gives your customer a chance to study your proposal apart from price considerations. Many contractors will ask you for a scope on sizable jobs ahead of time so that they can minimize the possibility of your excluding something for which they might later be held responsible. Because the scope proposal is submitted without the cost included, you should not sign it.

QUOTING PER PLANS AND SPECS

If you are quoting per plans and specs, this means that you will abide by the requirements of the specifications as they pertain to the work on which you are quoting and that you will also abide by any interpretations of them. This, of course, is exactly what your customer prefers, for it indicates that you are willing to base your price for the system on what is written, and further, on what may be construed by the architect in a dispute over the meaning of the specifications. When you know the architects and consulting engineers and have a good working relationship with your customer, there's no reason why you shouldn't be willing to quote the job per plans and specs. When this is the case, I think it is appropriate merely to type in the "Included" section of Form 20-1 the sentence, "This quotation is offered per plans and specifications," and not to bother listing items in the "Included" and "Excluded" sections. This tells your customer that you are comfortable with your interpretation of the requirements for the job as spelled out in your section of the specifications and that you will execute the job in accordance with the intent of the spec. To me, this is the best kind of quote, for numerous inclusions and exclusions can cloud the issue and cause your customer to doubt your ability to do the job as detailed by the contract documents for the price offered. The best quote is the simplest quote and one that is based on the construction documents as presented for bidding. There tends to be some hectic price manipulating among temperature control contractors at bid time, and so keeping your scope simple and complete is in your best interest. However, be sure that you understand plan and spec requirements.

QUOTING DIRECTLY WITH THE OWNER

When you are working directly with the owner of a building, especially when you are doing retrofit work, you might prefer to put your quote in a letter that completely details your proposal. In this case, there will be no plans or specifications, and instead your design and layout become the basis for the work.

Suffice it to say that whatever you offer to do, make sure that your offer is put in writing and that it is signed by both parties. Written agreements are the best agreements, for they can't forget what was said. And even if the original acceptor changes jobs, a written record of what was agreed upon remains.

Good luck to all of you. I hope you find that the work is as exciting and challenging to you as it has been to me.

APPENDIX

Table A-1 • Conversion Table for Units of Measure

1 inch (in) =	2.54 centimeters (cm)
1 foot (ft) =	30.48 centimeters (cm)
0.621 mile (mi) =	1 kilometer (km)
10 square feet (ft^2) =	0.9290 square meters (m^2)
10 cubic feet per minute (cfm) =	0.2832 cubic meters per minute (m^3/min)
8 hours per 100 feet (h per 100 ft) =	8 hours per 30.48 meters (h per 30.48 m)
2.6417 gallons (gal) =	10 liters (L)
1.0567 gallons per minute (gal/min) =	4 liters per minute (L/min)
10 pounds (lb) =	4.5359 kilograms (kg)
100 pounds per hour (lb/h) =	45.3592 kilograms per hour (kg/h)
14.2233 pounds per square inch (lb/in^2) =	1 kilogram per square centimeter (kg/cm^2)

Table A-2 • Temperature Conversion Formulas

Converting degrees Celsius to degrees Fahrenheit:

$$°F = \frac{°C}{100} \times 180 + 32$$

Converting degrees Fahrenheit to degrees Celsius:

$$°C = \frac{°F - 32}{180} \times 100$$

Table A-3 • Inches to Millimeters

Tubing, OD		Conduit (Trade Size)	
Inches	Millimeters	Inches	Millimeters
$\frac{5}{32}$	3.969	$\frac{1}{2}$	12.700
$\frac{3}{16}$	4.763	$\frac{3}{4}$	19.050
$\frac{1}{4}$	6.350	1	25.400
$\frac{3}{8}$	9.525	$1\frac{1}{4}$	31.750
$\frac{1}{2}$	12.700	$1\frac{1}{2}$	38.100
		2	50.800

Index

About the author

Harry J. Edwards, Jr., a controls contractor, licensed electrical contractor, and associate member of the International Association of Electrical Inspectors, has been in the controls industry for over 22 years, including 9 years with Robertshaw Controls and 10 with Honeywell. He now specializes in pneumatic controls systems service and reengineering for retrofit and better energy management. His first book, *Automatic Controls for Heating and Air Conditioning,* was published by McGraw-Hill in 1980.